Muscle Biology

A Series of Advances

VOLUME 1

Muscle Biology

A SERIES OF ADVANCES
VOLUME 1

Edited by

R. G. CASSENS

Muscle Biology Laboratory
College of Agricultural and Life Sciences
1805 Linden Drive
Madison, Wisconsin

and

THE INSTITUTE FOR MUSCLE BIOLOGY

MARCEL DEKKER, INC. New York   1972

MARCEL DEKKER, INC.
95 Madison Avenue, New York, New York 10016

LIBRARY OF CONGRESS CATALOG CARD NUMBER 72-81501

ISBN 0-8247-1092-4

PRINTED IN THE UNITED STATES OF AMERICA

PREFACE

This volume is the result of a lecture series sponsored by the Institute of Muscle Biology and conducted during the spring semester of 1971. The subject matter of the lecture series is based on a broad range of topics on muscle all of which represent recent major advances. The lecture series was organized in view of the rapid developments in the field and in consideration of the value of communication and discussion among scientists of the most recent work on muscle. The Institute of Muscle Biology was established to encourage the exchange of information and the development of interdisciplinary research and is composed of faculty members, from a number of departments, who have an interest in some aspect of muscle.

This volume is the result of cooperative effort of all members of the Institute of Muscle Biology; special thanks is expressed to the executive committee composed of E. Bittar, S. Kornguth, D. Slautterback, J. Will and R. Wolf for their encouragement and assistance. Appreciation is extended to E. J. Briskey who originally conceived and organized the lecture series. The willing cooperation of all lecturers is acknowledged.

Financial support was provided by the Graduate School of the University of Wisconsin-Madison, by the School of Medicine, by the College of Agricultural and Life Sciences, and by a grant-in-aid from the Campbell Soup Company. Contributing units include the Muscle Biology Laboratory, Molecular Biology Laboratory, Department of Anatomy, Cardiovascular Research Laboratory, Neurochemistry Program, Department of Preventive Medicine and Department of Pathology. Thanks is extended to M. FitzSimmons and J. Zwaska for secretarial assistance.

R.G.C.

Stephen A. Asiedu, Institute for Muscle Disease, New York, New York

Valerie Askanas, Institute for Muscle Disease, New York, New York

J. K. Barclay, University of Florida, Department of Physiology,
College of Medicine, Gainesville, Florida

Allan J. Brady, Department of Physiology and Los Angeles County
Cardiovascular Research Laboratory, University of California at
Los Angeles, Center for Health Sciences, Los Angeles, California

Bruce M. Carlson, Department of Anatomy, University of Michigan,
Ann Arbor, Michigan

W. King Engel, Medical Neurology Branch, National Institute of
Neurological Diseases and Stroke, National Institutes of Health,
Public Health Service, U.S. Department of Health, Education, and
Welfare, Bethesda, Maryland

Alfred L. Goldberg, Department of Physiology, Harvard Medical School,
Boston, Massachusetts

Richard W. Lymn, Department of Biophysics, University of Chicago,
Chicago, Illinois

Jean M. Marshall, Section of Neurosciences, Division of Biological
and Medical Sciences, Brown University, Providence, Rhode Island

Ade T. Milhorat, Institute for Muscle Disease, New York, New York

Wilfried F. H. M. Mommaerts, Professor and Chairman, Department of
Physiology and Director, the Los Angeles County Heart Association
Cardiovascular Research Laboratory, Professor of Medicine, Pro-
fessor of Molecular Biology, U.C.L.A. School of Medicine, Los
Angeles, California

W. K. Ovalle, The Neurophysiology Laboratory, Department of Surgery,
University of Alberta, Edmonton, Alberta, Canada

S. Ahmad Shafiq, Institute for Muscle Disease, New York, New York

R. S. Smith, The Neurophysiology Laboratory, Department of Surgery,
University of Alberta, Edmonton, Alberta, Canada

W. N. Stainsby, University of Florida, Department of Physiology,
    College of Medicine, Gainesville, Florida

Edwin W. Taylor, Department of Biophysics, University of Chicago,
    Chicago, Illinois

John R. Warmolts, Medical Neurology Branch, National Institute of
    Neurological Diseases and Stroke, National Institutes of Health,
    Public Health Service, U.S. Department of Health, Education, and
    Welfare, Bethesda, Maryland

CONTENTS

CHAPTER 1

ENERGETICS OF CONTRACTION

Wilfried F. H. M. Mommaerts

Professor and Chairman, Department of Physiology
and
Director, the Los Angeles County Heart Association
Cardiovascular Research Laboratory
Professor of Medicine, Professor of Molecular Biology
U.C.L.A. School of Medicine
Los Angeles, California

Approximately half a century ago, a major effort began to
formulate one of the central problems of physiological energetics:
What is the regulation of work performance and heat production and
what is the chemical origin of energy in muscular contraction?
Even though it was rooted in important older investigations, this
effort, which began with parallel investigations by Hill and by
Meyerhof [cf. 13], constitutes a concrete starting point for the set of
problems to which we are just now reaching the answers. During the
earlier phase of this study the entire development of our present
knowledge of cellular oxidation and glycolysis can be seen: the
concept of the generation of ATP as a result of metabolism and its
use as an energy source (and, specifically for muscle, phosphorylcreatine, or PC, as a ready reserve of ATP), and the major catego-
rization concerning the amount and time-distribution of the
mechanical and thermal energy, and associated metabolism, released
in connection with contractile activity [13]. More recently, the
major progress toward the elucidation of the ultrastructure of the
myofilament, of the actual workings of the constituent proteins,
and of the ultimate mechanism of the on-off control process was
added. These investigations assumed that a breakdown of ATP would

cover the required energetics, but only in the last decade has the actual use of ATP in the primary reaction coupled with activity been demonstrated, and only now, in the most recent years, are we finishing our balance of thermal, mechanical, and primary chemical events.

I would like to take this opportunity of the opening lecture in the series sponsored by the Muscle Biology Laboratory to present an outline of what is emerging in this field. Contemporary development is entirely dependent upon the hypothesis, now a decade old, that during a twitch or brief contraction cycle, the amount of chemical change that occurs is of the right order of magnitude to account for the energy turnover, the typical amount being set at about 0.3 µmoles chemical change per gram per cycle. This was first demonstrated by us (1960 onward) for PC [14,17,19], then by Davies et al. (1962 onward) for ATP [1,11]. Another conclusion-- going back to earlier workers, but mainly to Fenn, also about 50 years ago--is that, as suggested by heat measurements, the amount of this primary reaction depends on the mechanical conditions during the contraction [16]. For the correlation between work performance and metabolism over and beyond isometric contraction, the chemical counterpart of this was first obtained by Mommaerts, Seraydarian, and Maréchal [18] and by Carlson, Hardy, and Wilkie [2], and considerably extended by Maréchal [12] with the following result: If activity in the form of isometric contraction at stan- dard length l_o entails an amount of metabolism I, then it is found that in isotonic contractions the metabolism is about I + w, where w represents an amount of chemical change m_w such that its enthalpy change equals about 10,000 cal/mole, which is the total enthalpy effect of PC breakdown under the prevalent conditions of neutral- ization and buffering; w is, of course, the work performed. This is exactly the result that would correspond to the Fenn effect in the form it was originally published, and is close to what was found by us, including Maréchal. In the investigation of Carlson et al. [2], w and m_w by a certain count corresponded to about 6000

cal, and in experiments by Cain, Infante, and Davies [1] to even
much less than that. While it was not understood at that time,
these results were a first indication that the classical formulation of the Fenn effect did not cover all possibilities but represents merely one special case. The final result [16] must be
presented via a serious dilemma.

If we call E the energy turnover in a contraction cycle, such
as a twitch, and I the energy turnover in an isometric twitch at
l_o, we then have these situations (in which the energy quantities
E, I, w, etc., must be derived from chemical changes to m_E, m_I, m_w,
etc.):

Isometric twitch	$E = I$	
Fenn effect	$E = I + w$	(1)

No problems arose until Hill [6] demonstrated that heat production
accompanied the shortening process, and was proportional to it;
this factor became known as the shortening heat

$$E = I + a(-\Delta L) + w \qquad (2)$$

and, while we can see the discrepancy immediately, we must make
clear just what realities these two equations represent.

Equation (1) represents the following. We imagine a muscle
(on a thermopile) at l_o that upon single stimulation delivers the
isometric twitch heat as defined. When it is instead made to lift
a load P over a distance $(-\Delta L)$, again starting at l_o, the muscle
does work, $w = P(-\Delta L)$. The measurement, in the typical Fenn experiment, shows that it also develops the same heat I (if we allow the
lifted load to fall back upon the muscle, it turns out that w is
returned to the muscle as heat, so that E can be obtained thermally,
but that is a side issue), hence Eq. (1). In a way, this is unexpected, because in the form presented it suggests that the muscle,
carrying out its overhead I as usual, in addition mobilizes metabolism to cover the work it does, and performs this work with
perfect enthalpy efficiency--a most unlikely situation. The facts
underlying Eq. (2) actually show the opposite: when the muscle

shortens, it delivers work and heat at the same time:

$$E = I + a(-\Delta L) + P(-\Delta L)$$
$$E = I + (a + P)\ (-\Delta L) \tag{3}$$

Thus we can only conclude that the factors I in the two equations
cannot possibly be the same, as was first recognized by Mommaerts,
Seraydarian and Maréchal [18].   The point is, that in the transition from Eq. (1) to (2) or (3), I diminishes to about the extent
that a(-ΔL) appears; but that the original shortening heat experiments of Hill [6] were so staged that this circumstance did not
attract attention.   The solution finally reached [15,16] is that I
is a composite factor, consisting of A, which is the energy turnover connected with switching the muscle on and off, and of f(P,t)
related to the presence and maintenance of tension.   By definition,
A is constant in the context, though we must be aware that the
intensity of excitation-contraction coupling changes, e.g., in the
inotropic changes of cardiac muscle, or that e-c coupling diminishes
upon extreme shortening, according to recent work by Rüdel and
Taylor [21].   A will be called the activation heat.

$$E = A + f(P,t) + a(-\Delta L) + w \tag{4}$$
$$I = A + f(P,t) \tag{5}$$

Since (-ΔL) becomes larger the smaller the load, it will be seen
that f(P,t) and a(-ΔL) change in a roughly reciprocal manner.   In
some muscles, and this appears to be the case for some frog muscles
at 0°C, these opposite changes seem to be approximately equal,
which explains Eq. (1) and the original Fenn effect.   A new approach
to the measurement of the shortening heat, first used by Gibbs,
Mommaerts, and Ricchiuti [4] illustrates this in a most direct
manner, and was developed with satisfactory precision by Homsher
et al. [8,9].   The coefficient a in Eqs. (3)-(4) is not constant
but is a moderately weak function of P [7].   Its value and variation, measured with the new method, are the same as with the Hill
procedure [9].

    It is important to realize that a heat of shortening is not

only an experimental finding, but also a theoretical necessity.
If, after appropriate subtraction of a reference baseline, no heat
were to accompany the shortening process, this would be the same as
stating that the enthalpy efficiency during work performance would
be unity. Thus,

$$\alpha = \frac{w}{m_w \cdot (-\Delta H)}$$

$\alpha = 1$ (in the supposition examined)

$$m_w (-\Delta H) = m_w T (-\Delta S) + m_w (-\Delta F)$$

$$\varepsilon = \frac{w}{m_w \cdot (-\Delta F)}$$

$$w = m_w \cdot T(-\Delta S) + w_\varepsilon$$

$$\varepsilon = \frac{w}{w - m_w \cdot T(-\Delta S)}$$

$$\varepsilon = \frac{1}{1 - (T \cdot \Delta S)/(\Delta H)}$$

(6)

This may happen to be true at one given speed, but not at all at
other speeds because certainly ε must diminish with increasing
shortening velocity. Another derivation [15, pp. 476-477] shows
that, apart again from an undefined baseline, the isotonic shortening heat is proportional to the distance shortened, which is indeed
its principal experimental characteristic.

The remaining major problem, then, is that of the chemical
origin of the various terms in Eq. (4). While we have accepted,
for the present discourse, that in the first approximation the
energy in a twitch is covered by the use of either ATP or PC, we
must keep an open mind about this concerning the details, and not
exclude from scrutiny whether a relatively small fraction might not
come from something else. As a case in point, it might be mentioned
that Ricchiuti and I once found that the isometric heat in a greatly
shortened muscle (approximately A at a value of about 0.001 cal/g)
was not followed by an equally large restitution heat. We entertained the hypothesis that A might not come from ATP but from some

one of the intermediates discussed by the workers in the field of
oxidative phosphorylation. We now have found differently, and
assume that the small restitution heat is the consequence of the
limited response of the metabolic control system to a very small
signal; but the episode deserves mention as an example of the
specific issues that can only be decided by actual tests, not by
the conviction that, by and large, the source of energy was covered.
This may be the occasion to discuss to what extent ATP and PC
deserve equal emphasis, or whether all our attention should be
directed toward ATP as the primary action substance, and PC should
be dismissed as a secondary phenomenon. I believe that both deserve
equal emphasis in this context, because factors such as the actual
$(-\Delta F)$ will be determined by the PC-products concentration ratio as
much as by [ATP]/[ADP], and are surely buffered by the former, not
the latter. Also experimental considerations speak for this, since
the aerobic, the anaerobic, the iodoacetate-poisoned, and the
fluorodinitrobenzene-poisoned muscle each have their own experimental peculiarities, advantages, and characteristics, but surely
become unhealthier in the order of listing.

As to the actual chemical sources of the various energy terms,
we can express certain anticipations, which must be experimentally
confirmed if our standpoint is correct. First of all, w and $A(-\Delta L)$
must come from the same chemical source; this follows from the
entire preceding argument as follows, leaving out the terms that
deal with A and $f(P,t)$:

$$
\begin{aligned}
E_w &= m_w \cdot T(-\Delta S) + m_w \ (-\Delta F) \\
    &= m_w^* \cdot T(-\Delta S) + m_w \ (1 - \varepsilon)(-\Delta F) + m_w \cdot \varepsilon \ (-\Delta F) \\
    &= \underbrace{m_w \ [T(-\Delta S) + (1 - \varepsilon)(-\Delta F)]}_{} + \underbrace{m_w \cdot \varepsilon \ (-\Delta F)}_{} \qquad (7) \\
    &= \qquad\quad \text{shortening heat} \quad + \quad\ \text{work}
\end{aligned}
$$

Thus, if work comes from ATP breakdown, the shortening heat must
too; any hypothesis to the contrary would violate elementary
principles.

It is not analytically demonstrable, from first principles only, that $f(P,t)$ stems from the same chemical source as well. It becomes, however, extremely likely that this is so when we consider the basic mechanism of chemomechanical transduction, and it seems quite inherent in the reigning theory, that of Huxley [10] (see Mommaerts [15] for its relation to the current view of the cross-bridges). Thus, our current standpoint, if correct, would require all three chemomechanical terms to be equivalent to amounts of ATP or PC broken down.

This is not necessarily so--or at least not for the same reasons--for A, which is the net heat that falls by the way in the course of a complete cycle of excitation-contraction coupling. We can study A separately by stretching the muscle to such a length that the overlap of A-band and I-filaments is abolished so that no interacting crossbridges remain [9,22]. Without thereby exhausting the subject, let us say that A is the algebraic sum of some events such as the following: (i) the enthalpy effect of the liberation of sarcoreticular calcium ions from their binding or storage form, and including the heat of dilution; (ii) the heat of their combination with troponin-tropomyosin, including whatever conformational change this may induce in these proteins or in actin, and including events such as the liberation of H^+ ions from troponin and their neutralization; (iii) the removal of Ca^{2+} ions from troponin and the reversal of the accompanying reactions; (iv) the resequestering of Ca^{2+} ions back into their sarcoreticular-to-storage sites; (v) the enthalpy effect of whatever metabolic reactions accompanies one or the other of these events, and is linked with it as a driving reaction. This listing, which is likely to be oversimplified, looks formidable, but can actually be reduced to one item when we concern ourselves only with the total energetic change over a whole contraction cycle. If the muscle would be in the same state afterward as before, it would be evident that items (i) and (iv), and (ii) and (iii) cancel completely. (There is some question whether, under limited experimental conditions, the Ca^{2+} ions return exactly

to where they came from, but this can be disregarded if we make our
comparisons appropriately for comparably chosen steady-state or
transient conditions.) It then follows that the total end value of
the activation heat must come from the enthalpy of its driving
reaction (v).

According to our current results [8] this driving reaction is,
again, a splitting of ATP, and in a steady state a breakdown of 0.1
μmole corresponds to an activation heat of 1 mcal. This does not
at all follow from the previous argument linking $f(P,t)$, $a(-\Delta L)$,
and w, with which it is not at all related. It is not, however,
unexpected in view of the existing knowledge linking ATP with Ca^{2+}
pumping by sarcoreticular vesicles [3,5]. To further clarify our
views on items (i) through (v) it would be necessary to have more
information about the time sequence of events. Actually, we have
learned that the activation heat occurs in two phases: a rapid
initial one with a time constant of the order of 1/20 sec which is
not sensitive to temperature, and a highly temperature dependent
phase which at 0° has a time constant of 1/3 sec, but which at room
temperature cannot be resolved from the former. The most obvious
explanation would be to ascribe the rapid phase to item (i) or to
(i) and (ii) jointly. The slow phase then represents the heat of
ATP-splitting minus the enthalpy effects of the reversals of (i)
and (ii). In the light of this temporal resolution, it is of
interest to reexamine an old controversy about the demonstrable
time course of ATP breakdown in a twitch. Infante and Davies [11]
stated in 1962 that at the peak of a twitch, 0.22 μmole^{-1} of ATP
is used, as against 0.43 μmole g^{-1} halfway through relaxation. In
1967, Mommaerts and Wallner [20] denied this, showing an ATP use of
about 0.3 μmole g^{-1} for the cycle, and during the entire relaxation
phase no use of more than a few hundredths of a micromole.

What can be expected? Firstly, the delayed ATP breakdown cannot on this basis amount to 0.2 μmole. Secondly, it cannot, for
the most part, occur during the mechanical relaxation phase since
this, by any analysis (compare with, e.g., Huxley [10]), lags

behind the actual inactivation mechanism. We predict that the
breakdown of ATP related to excitation-contraction coupling starts
shortly or immediately after the rapid phase of the activation heat
inasmuch as the resequestering of Ca^{2+} is likely to start the moment
the ejection phase is finished, and inasmuch as the ATP breakdown
is to a sizable degree complete by the time mechanical relaxation
past the twitch peak becomes evident. Thus, the picture is in
agreement with the results of Mommaerts and Wallner [20], which
would allow the tailing of ATP breakdown by a few hundredths of a
micromole or so, but not the much larger amount which in any case
is in excess of anything relatable to calcium pumping.

The chemical identification of the other items in Eq. (4) has
now also been achieved quite satisfactorily [8,9]. The difficulty
was that, after fixing A, there are still three independent variables and, of course, limited accuracy inherent in the chemical
analyses. On the other hand, the Gibbs, Mommaerts, and Ricchiuti
[4] myothermal procedure provided guidance. Thus, unlike in the
earlier investigation of Mommaerts, Seraydarian, and Maréchal [18],
and of Carlson, Hardy, and Wilkie [2], we could explicitly segregate
certain factors (i) f(P,t) by measuring myothermal and biochemical
events in isometric contractions of muscles in which P was varied
by stretch so as to reduce the number of crossbridges, (ii) a(-ΔL)
by comparing (by extrapolation) the heat and chemistry of unloaded
shortenings with those in zero-force isometric contractions. The
results will not be given here in further detail since the publications are in preparation. It will suffice that we have achieved
a biochemical accounting for the entire equation [4], with satisfactory precision, except that the biochemical change for the
shortening heat at the moment of writing is somewhat less precise
than the other factors which may be acceptable within the error
margin (considering what we are doing), or which may still be traced
to a small difference between the instrumentations for the myothermal and the biochemical experiments.

At the time of the comprehensive review [15] (1969), I stated

that there were two major problems left: the activation heat was
only partly accounted for, and the shortening heat not at all. As
a result of our work of the last few years, this gap has now been
closed by the finding that all events in activated muscles are
accounted for by the primary reaction of ATP and PC breakdown. We
can now turn our attention to the next half-century, and express
the wish that the Muscle Biology Laboratory will be a vital parti-
cipant in it.

REFERENCES

[1]   Cain, D. F., A. A. Infante, and R. E. Davies, "Chemistry of
muscle contraction. Adenosinetriphosphate and phosphorylcreatine
as energy supplies for single contractions of working muscle,"
Nature, 146, 214-217 (1962).

[2]   Carlson, F. D., D. J. Hardy, and D. R. Wilkie, "Total energy
production and phosphocreatine hydrolysis in the isotonic twitch,"
J. Gen. Physiol., 46, 851-882 (1963).

[3]   Ebashi, S., and M. Endo, "Calcium ion and muscular contraction," Progr. Biophys. Mol. Biol., 18, 125-183 (1968).

[4]   Gibbs, C. L., W. F. H. M. Mommaerts, and N. V. Ricchiuti,
"Energetics of cardiac contraction," J. Physiol. (London), 191, 1,
25-26 (1967).

[5]   Hasselbach, W., "Relaxing factor and the relaxation of
muscle," Progr. Biophys., 14, 169-222 (1964).

[6]   Hill, A. V., "The heat of shortening and the dynamic constants
of muscle," Proc. Roy. Soc. (London), Ser. B, 126, 136-195 (1938).

[7]   Hill, A. V., "The effect of load on the heat of shortening of
muscle," Proc. Roy. Soc. (London), Ser. B, 159, 297-318 (1964).

[8]   Homsher, E., W. F. H. M. Mommaerts, N. V. Ricchiuti, and A.
Wallner, "Activation heat, activation metabolism, and tension-
related heat in frog semitendinosus muscles," J. Physiol. (in
press, 1972).

[9]   Homsher, E., W. F. H. M. Mommaerts, N. V. Ricchiuti, and A. Wallner, "Shortening heat and work and their chemical origins" (paper in preparation, 1972).

[10]   Huxley, A. F., "Muscle structure and theories of contraction," Progr. Biophys. Biophys. Chem., 7, 257-318 (1957).

[11]   Infante, A. A., and R. E. Davies, "Adenosinetriphosphate breakdown during a single isotonic twitch of frog sartorius muscle," Biochem. Biophys. Res. Commun., 9, 410-415 (1962).

[12]   Maréchal, G., Le metabolisme de la phosphorylcreatine et de l'adenosine triphosphate durant la contraction musculaire, Editions Arscia, Brussels, 1964, p. 1-184.

[13]   Meyerhof, O., Die chemischen Vorgänge im Muskel und ihr Zusammenhang mit Arbeitleistung und Wärmebildung, Springer-Verlag, Berlin, 1930, p. 1-350.

[14]   Mommaerts, W. F. H. M., "The regulation of metabolism and energy release in contracting muscle," Circulation, 24, 410-415 (1961).

[15]   Mommaerts, W. F. H. M., "Energetics of muscular contraction," Physiol. Revs., 49, 427-508 (1969).

[16]   Mommaerts, W. F. H. M., "What is the Fenn-effect?" Naturwiss., 57, 326-330 (1970).

[17]   Mommaerts, W. F. H. M., M. Olmsted, K. Seraydarian, and A. Wallner, "Contraction with and without demonstrable splitting of energy-rich phosphate in turtle muscle," Biochem. Biophys. Acta, 63, 82-92 (1962).

[18]   Mommaerts, W. F. H. M., K. Seraydarian, and G. Maréchal, "Work and chemical change in isotonic muscular contraction," Biochim. Biophys. Acta, 57, 1-12 (1962).

[19]   Mommaerts, W. F. H. M., K. Seraydarian, and A. Wallner, "Demonstration of phosphocreatine splitting as an early reaction in contracting frog sartorius muscle," Biochim. Biophys. Acta, 63, 75-81 (1962).

[20]   Mommaerts, W. F. H. M. and A. Wallner, "The breakdown of
adenosine triphosphate in the contraction cycle of the frog sartorius muscle," J. Physiol. (London), 170, 343-357 (1967).

[21]   Rüdel, R. and S. R. Taylor, "Striated muscle fibers: Inactivation of contraction induced by shortening," Science, 167,
882-884 (1970).

[22]   Smith, I. C. H., "Energetics of activation in frog and toad
muscle," J. Physiol., (in press, 1972).

CHAPTER 2

CELL AND TISSUE INTERACTIONS IN REGENERATING MUSCLES

Bruce M. Carlson

Department of Anatomy
University of Michigan
Ann Arbor, Michigan

13

The regeneration of a limb muscle represents a complex
developmental event which involves both the cytodifferentiation of
individual muscle fibers and the organization of these muscle fibers
into a grossly identifiable structure. These two processes occur
simultaneously, and they do not take place as anatomically or
functionally isolated events. Rather, they are often dependent
upon interactions with other components of the limb. In this brief
review I intend to discuss two experimental means by which entire
limb muscles can be regenerated. One is the regeneration of muscles
from minced fragments (tissue regeneration) and the other is the
regeneration of muscles within a regenerating limb (epimorphic
regeneration). In describing the process of regeneration in each,
I shall emphasize not only well demonstrated tissue interactions,
but also those areas in which a lack of information about basic
elements of the regenerative process precludes an understanding of
regulatory or interactive events.

I. MINCED MUSCLE REGENERATION

The regeneration of entire muscles from minced fragments was
first described by Studitsky [59,61] and many parameters of this
experimental system have subsequently been investigated in his
laboratory [43,64,81]. The method of producing minced muscle
regeneration consists of entirely removing a muscle, mincing it
into fine fragments (about 1 mm^3) and implanting the muscle
fragments into the original muscle bed [4,66]. The degenerating
implanted muscle fragments give rise to a population of myogenic
cells which, developing along with connective tissue, blood vessels,
and nerves, give rise to a model of the original muscle. The
regenerated muscle is always considerably smaller than the original
one, but the general anatomical connections and some functional
properties are restored. Minced muscle regeneration has been
obtained in axolotls [5] and newts [Carlson and Hsu, unpublished],
in birds [60], and in several species of mammals [4,15,62,77,81].
A typical minced muscle regenerate in the rat is illustrated in
Fig. 1.

Figure 1. Somewhat smaller than average minced muscle regenerate in the rat, two months post-operatively.

A. Effects of Mincing

What is done to the muscle by the mincing process, and how might these effects be related to the subsequent regenerative process?

1. The muscle has been subjected to severe mechanical trauma which has disrupted the structural integrity of the muscle fibers. It has been demonstrated that severely damaged muscle fibers do not normally undergo a local repair process of the damaged elements, but instead they degenerate and subsequently give rise to new muscle fibers [2].

2. The minced muscle has been denervated. The effects of denervation upon intact adult muscle have been extensively studied

and often reviewed [18-20], but less attention has been paid to
the relationships between denervation and regenerative phenomena.
Studitsky [62], has long maintained that denervation (as well as
some other experimental manipulations) causes muscle to enter into
a "plastic" state, in which condition the ability of muscle to
undergo restorative processes [58] and to survive auto- or homotransplantation [65] is considerably improved. Yeasting [77] has
recently noted that denervation of mouse muscle 2 or 4 weeks prior
to severe trauma results in an acceleration of the early phases of
regeneration. Recent morphological studies by Lee [30] and Hess
and Rosner [24] provide at least a clue to possible relationships
between denervation and subsequent regenerative activity. These
authors found a marked increase in the number of satellite cells
under the basement membranes of denervated muscle fibers in rats,
rabbits and guinea pigs and presented evidence indicating that
denervation causes myonuclei to pinch off from the muscle fibers
themselves and to collect under the basement membranes as a potential population of progenitor cells of new muscle.

     3. The muscle has been separated from its vascular supply.
It has been known for years [31,32] that periods of ischemia produce
a necrotic reaction followed by a marked regenerative response in
skeletal muscle. Electron microscopic studies by Reznik [44] have
demonstrated that ischemia of mammalian skeletal muscle is quickly
followed up by the establishment of a population of myogenic cells.

     4. Intracellular muscle proteins are released into the general
circulation. The significance of this in relationship to regeneration is unknown. Except for clinical studies relating to myasthenia
gravis, very little work has been conducted on the antigenic properties of injured muscle or autologous muscle proteins. Since muscle
regeneration can occur in animals which, due to their phylogenetic
or ontogenetic development, have poorly developed immune responses,
it is doubtful whether cellular or humoral immune responses play a
critical role in this process. However, the appearance of light
infiltrations of mononuclear leukocytes during certain stages of

regeneration suggests that the lymphoid system of the animal may
not be totally unresponsive to the injury inflicted upon the muscle.
These mononuclear cells are characterized by slightly greater
amounts of cytoplasm than those which are seen in the classical
homograft rejection response to foreign muscle tissue [6,63].

5.   The muscle fibers are released from tension. Normal
muscle fibers are constantly exposed to a certain degree of tension,
even at rest, and it is known that tenotomized muscles undergo a
characteristic series of degenerative changes [25]. Le Gros Clark
and Wajda [33] commented upon regeneration following ischemic
necrosis in the tenotomized tibialis anterior muscle in rabbits.
According to their observations, the removal of degenerated tissue
and the growth of young muscle fibers was actually somewhat accelerated in comparison with nontenotomized controls even though later
degenerative changes were noted.

6.   The internal architecture and gross form of muscles are
totally disrupted by the mincing process. Although of no apparent
import with respect to the initiation of regeneration, it must be
noted that beyond the confines of any given muscle fragment, the
minced muscle contains no preformed scaffolding upon which morphogenesis can proceed.

Thus the mincing operation itself incorporates many of the
standard techniques which are used to stimulate regeneration in
other experimental systems. More important with respect to the
theme of this discussion is the fact that in a freshly implanted
minced muscle almost all of the normal anatomical, functional and
environmental relationships which a normal muscle enjoys have been
destroyed. The rest of the description of this process will be
devoted to a consideration of how these disrupted interrelationships
are restored.

B. The Degenerative Phase

In the immediate post-implantation period, the muscle fragments
undergo a widespread degenerative response. This is largely

independent of interactions with other components of the limb.
During this period, however, the muscle fibers retain certain properties which enable myogenic elements to survive the period of
degeneration and to make preparatory steps for the progressive phase
of the regenerative process. Metabolically, the degenerating muscle
fibers retain a surprising level of activity or, at least, potential
activity. In my laboratory Snow [56,57, unpublished] has been
investigating certain metabolic parameters of degenerating minced
muscle in the rat with histochemical and biochemical techniques.
In some areas the minced tissue may be devoid of a direct blood
supply for over a week. Snow finds that glycogen levels are reduced
rather quickly and that phosphorylase activity disappears within a
day. Histochemically demonstrable activity of the mitochondrial
enzymes, succinic dehydrogenase and cytochrome oxidase, is normal
during the first few hours after mincing. Succinic dehydrogenase
activity remains at high levels for several days before declining
whereas cytochrome oxidase activity steadily disappears at a more
rapid rate. Lactic dehydrogenase is both histochemically and bio-
chemically demonstrable throughout the entire period of muscle
degeneration. An increase in nonspecific esterase activity occurs
in the degenerating muscle midway through the first week after
mincing and neutral lipids are found in moderate amounts throughout
the degenerative period. These studies by Snow provide evidence
that, given the presence of substrates, a considerable degree of
metabolic activity, particularly anaerobic, is possible in the
degenerating muscle and that chemically, it is not a dead or necessarily inert system.

    Structural evidence lends support to this conclusion. Although
obvious sarcoplasmic degeneration occurs and a number of the nuclei
also break up, other nuclei retain a viable appearance as seen by
both light and electron microscopic techniques. Trupin [74, un-
published] has observed numerous healthy-appearing nuclei under the
basement membranes of degenerating minced muscle fibers in the frog.
Some of these have been connected to the underlying sarcoplasm by

cytoplasmic bridges. Autoradiographic studies [9] at the periphery
of degenerating muscle fibers in the frog can incorporate H^3-
thymidine. These studies were done at the light microscopic level
and cell types within the basement membrane could not be distinguished. Considerable incorporation of H^3-uridine has also been
observed around these nuclei.

Thus, during the phase of degeneration, elements of the minced
muscle tissue appear to retain certain metabolic pathways, and a
population of viable (presumably myogenic) cells becomes established
under the basement membranes of the degenerating muscle fibers.

C. Re-establishment of Relations with the Vascular Supply

The initiation of histologically obvious regeneration of muscle
is both spatially and temporally very closely associated with the
re-establishment of a vascular supply to the implanted muscle fragments. Vascular injection studies in rats [10] have shown that
blood vessels begin to grow into the surface of the muscle mass two
days after implantation. As the vasculature centripetally invades
the minced muscle, the regeneration of new muscle fibers occurs.
Soon three distinct zones of activity are established [5,10]: (a) A
central zone containing the originally implanted muscle fibers.
This zone is not vascularized and presents little histological evidence of regenerative activity. (b) A *peripheral zone* of intense
regenerative activity. This zone is fully vascularized and contains
multinucleated forms of regenerating muscle as well as regenerating
connective tissue. There is a particularly close association between
newly regenerating muscle fibers and the vascular network. There
are no traces of the original muscle fragments in this zone. (c)
Between the central and peripheral zone lies the *transitional zone*.
It is in this zone where the myoblasts enlarge and fuse to become
basophilic cuffs around the fragmenting sarcoplasm of the original
muscle fibers. The middle of the transitional zone also represents
the most inward penetration of the vascular supply. Within this
zone the activation of myogenic nuclei normally occurs a couple

hundred microns centrally from the furthest inward incursion of the
vasculature. In the early regenerate the peripheral zone represents
only a thin outer shell, and the bulk of the tissue is included in
the central zone. During succeeding days as the vasculature pene-
trates toward the center of the muscle mass, the peripheral zone of
regeneration increases in thickness at the expense of the shrinking
central zone. The transitional zone always retains its relative
position between the other two zones. Normally by the time 8-10
days elapse the old muscle fragments have completely degenerated
and the pattern of zonation disappears.

Although there seems to be a causal relationship between the
ingrowing vascular supply and the initiation of the progressive
phase of regeneration, this has not been unequivocally proven. It,
nevertheless, seems probable. Although biochemical work on this
system has been minimal, it is likely that the returning vascular
supply causes some rather profound metabolic shifts--probably from
an anaerobic to an aerobic type--and the supply of substrates for
energy production as well as a supply of precursor molecules for
macromolecular synthesis is probably considerably increased.

Another function of the invading vascular supply may be to
bring in phagocytic cells which remove most of the sarcoplasm of
the original muscle fibers. The regeneration of new muscle fibers
is very limited until massive phagocytosis of old sarcoplasm takes
place, and Allbrook [1] and Allbrook, Baker, and Kirkaldy-Willis [2]
have placed great stress on the importance of phagocytosis as a
prerequisite for muscle regeneration. A close parallel to the
observations on minced muscle regeneration is the reaction of pieces
of rat muscle placed in Millipore diffusion chambers [Carlson, un-
published]. Separated from both a direct blood supply and blood-
borne phagocytes, the nuclei of the muscle fibers undergo the
initial stages of activation. The old sarcoplasm, however, remains
essentially intact, and regeneration is arrested at a very early
phase for days. It would be interesting to test the effect of
introducing macrophages from peritoneal exudates into these

diffusion chambers upon the degree of destruction of old sarcoplasm and the regeneration of new muscle fibers.

D. Nerve-Muscle Relationships

About a week after implantation, depending upon the species, nerve fibers begin to find their way back into·the regenerate. How do nerves and muscle interact?

Initially, nerve fibers must penetrate into the substance of the regenerate. Our understanding of causal events related to this phenomenon is so limited that it cannot be said whether the regenerating nerve fibers are guided into the regenerate by factors in the physical environment or whether the regenerating tissue positively attracts the nerves in some other manner. Experiments similar to those conducted in embryonic systems are certainly indicated.

Whatever the reasons, nerves do return into the regenerate. A most important question relates to the effect of the ingrowing nerve supply upon the muscle regenerate. Hsu [28] has carefully investigated the re-establishment of innervation to the regenerating minced gastrocnemius muscle in frogs. Although the first nerve fibers penetrate into the regenerate by 9 days, they do not become associated with muscle fibers until 12-14 days. By this time many of the muscle fibers in the outer portions of the regenerates have already developed cross striations and peripherally located nuclei whereas those muscle fibers toward the center are less mature. Motor end plates are not demonstrable until at least two months in the frog, but eventually the regenerated muscles are capable of some contractile activity after stimulation of the sciatic nerve. In mice and in rats, contractile activity of minced muscle regenerates is established considerably sooner [77,78].

In order to shed light upon the role of the nerve in minced muscle regeneration, several investigators have studied the course of regeneration in limbs denervated at the time of operation. Working on the rat, Zhenevskaya [78] found that following denervation, regeneration of muscle fibers was delayed at the myoblastic

and early myosymplastic (myotubal) stages and that subsequent differentiation proceeded very slowly. However, histological observations in the experiment were not made before 10 days post-
operatively according to the experimental protocol. As a result of
this experiment, it was postulated [62,67] that whereas the earliest
stages of muscle regeneration could occur in the absence of nerves,
some type of trophic effect was required to bring about the critical
stages of maturation from the early multinucleated stage to a mature
muscle fiber. In further work Zhenevskaya [79] eliminated the motor
supply to the minced gastrocnemius muscle of the rat and noted that
profound atrophic changes affected the regenerative process even
though the sensory and sympathetic nerve supply to the regenerate
was abundant. In contrast, elimination of the sensory nerve supply
to regenerating muscles retarded the differentiative process to some
extent, but did not inhibit it [80]. Yeasting [77] followed the
regeneration of denervated minced muscles in the mouse and found
that early regeneration, including the differentiation of cross-
striated muscle fibers, was not delayed but that atrophic changes
occurred at a later period.

In the frog, Hsu [28] has examined numerous regenerating minced
muscles in denervated limbs at daily intervals and was able to
detect no histological difference between normal and denervated
regenerates until the end of the second week. Then some of the
regenerating muscle fibers became vacuolated, and their nuclei
underwent clumping and pycnotic changes. From this time there were
always fewer muscle fibers in denervated regenerates than in normally
innervated controls. By 30 days most of the muscle fibers had disappeared from denervated regenerates. Nevertheless, individual
muscle fibers did regenerate to full maturity in the absence of
nerves.

In summary, we can say the following about nerve-muscle interactions: (a) We do not know what attracts nerve fibers to the
regenerating muscle. (b) A nerve supply is necessary for full
regeneration of minced muscles. Nerves do not significantly affect

the early stages of regeneration, but a nerve supply is required
for the maintenance of later stages. Studitsky, Zhenevskaya and
Rumyantseva [67] think that nerves, by a trophic influence, control
the differentiation and growth of regenerating muscle fibers. Hsu
[28], on the other hand, concludes that the effect of an absent
nerve supply is felt at a somewhat later stage of development and
that the lack of nerves is reflected in a degeneration of already
differentiated muscle fibers as well as in a retardation of development of less mature fibers. (c) The trophic influence of nerves in
minced muscle regeneration appears to be mediated by the motor
fibers. This is in marked contrast to the quantitative rather than
qualitative nerve requirements in the regenerating amphibian limb
[53].

E. Interactions Involved in the Morphogenesis of Minced Muscle Regenerates

1. *The Role of Tendon Connections*

An early structural interaction is the establishment of fibrous
connections between the proximal and distal ends of the minced
muscle mass and the tendon stumps left behind at the original operation. Under normal circumstances the regenerating muscle retains
an elongated configuration during the course of regeneration.
Several experiments conducted upon rats have demonstrated that the
maintenance of an elongated form in the regenerate is dependent
upon the continuance of proximal and distal tendon connections. If
the distal tendon connection of the minced gastrocnemius muscle is
eliminated by amputation of the foot and complete removal of the
Achilles tendon stump, then the regenerate tends to assume a somewhat globular shape, and it extends only about one-third the length
of the limb segments [8,12,47]. If minced muscle is freely implanted
under the abdominal skin, the regenerating tissue rather quickly
rounds up into a button-like shape [8,12,46]. In contrast to this,
when the minced gastrocnemius muscle is implanted into a leg with
the foot amputated but with the stump of the Achilles tendon left

behind, the regenerate connects to the tendon stump and maintains
an elongated shape [12].

The results of a number of control experiments [Carlson, un-
published] have demonstrated that the connections between tendon
stumps and the minced muscle consist primarily of cells and their
fibrous secretions, which originated in the tendon stumps rather
than in the minced muscle. In the absence of minced muscle, the
outgrowing connective tissue makes contact with the underlying
tissues and forms firm adhesions.

2.   *The Role of Tension in the Establishment of Internal Architecture*

When multinucleated muscle fibers first appear in minced muscle
regenerates, they are oriented almost randomly within the overall
regenerating mass. As time progresses, the myotubes and muscle
fibers become oriented parallel to the long axis of the limb. This
reorientation first occurs in the periphery and works its way toward
the center. In rats, a high percentage of muscle fibers will be
correctly oriented by the third week of development. Internal
organization, however, is seldom as regular as it is in normal
muscle.

Earlier experimentation by Rumyantseva [46,47] indicated that
tension is an important factor in the establishment of a normal
internal organization of minced muscle regenerates. She also
stressed the importance of tension as a factor in the cytodifferentiation of striated muscle fibers. Several experiments on rats
have demonstrated that a decrease in tension is associated with a
lack of internal organization in the regenerates. Immobilization
of the hind limb in a plaster cast results in some irregularity in
internal architecture, but overall orientation of muscle fibers
still occurs [8]. As was mentioned above, amputation of the foot
and removal of the stump of the Achilles tendon results in a roughly
globular regenerate. Internally, considerable disorientation of the
muscle fibers occurs [12,47], but in older regenerates bands of well

oriented muscle fibers can be seen. This is likely due to a small
degree of tension exerted by the lower limb segment upon the origin
of the regenerated muscle on the femur. Muscle fragments freely
implanted beneath the abdominal skin exhibit a complete three
dimensional disorganization in the orientation of muscle fibers [8,
12,46]. In contrast to the findings of Rumyansteva, regenerating
muscle fibers in my abdominal implants developed cross striations
and peripheral nuclei before degenerating, despite a lack of tension
and minimal innervation. From these experiments it can be concluded
that the establishment of a regular internal architecture in minced
muscle regenerates does not occur in the absence of tension.

Recent experiments [12] have shown that when tension is re-
applied to a formerly tension-free system, the irregular orientation
of muscle fibers can be corrected. When the Achilles tendon was
left behind after amputation of the foot, the orientation of muscle
fibers in the regenerated gastrocnemius muscle approached, but did
not quite attain, that of a normal regenerate. An experiment
conducted upon minced muscle implanted beneath the abdominal skin
emphasizes the role of tension in the internal organization of
minced muscle regenerates. The gastrocnemius muscle was minced and
implanted under the abdominal skin in the usual fashion. At both
the anterior and posterior poles of the implanted muscle fragments,
a segment of Achilles tendon was positioned. Each tendon was connected by sutures to skeletal elements (the anterior one to a rib
and the posterior one to a public ramus), which represented fixed
points on the elongating body. Within a few days the tendons made
connections with the muscle fragments, and as the rats grew in
length, the tendons were pulled away from each other. The regenerating minced muscle was in the direct line of tension, and instead
of rounding up, it became considerably elongated. Internally
approximately three-quarters of the muscle fibers became oriented
parallel to the lines of tension. The fibers at the edges of these
regenerates tended to remain somewhat disorganized.

3.   The Role of Internal Connective Tissue

Although tension is obviously applied to minced muscle regenerates through the proximal and distal tendons, this does not
explain the propagation of tension within a regenerate. Incidental
findings in an experiment by Gallucci, Novello, Margreth, and Aloisi
[15] indicate that the connective tissues within a regenerating
muscle serve a very important role in orienting the muscle fibers.
They performed minced muscle operations on rats which had been made
lathyritic by the administration of aminoacetonitrile. They stated
that in contrast to regenerates in normal animals, the internal
architecture in connective tissue-poor regenerates was quite irregular.

4.   The Role of the Surrounding Tissues in the Attainment of Gross
     Morphogenesis

One of the earliest felt morphogenetic forces is merely the
physical pressure of the muscles and skin surrounding the implanted
muscle fragments. This pressure keeps the fragments in a roughly
cylindrical shape until some internal coherence develops within the
muscle mass. In control experiment designed for other purposes
[Carlson, unpublished], cell free homogenates of muscle were soaked
into pieces of Gelfoam, which were then implanted in place of the
gastrocnemius muscle. When the rats were autopsied 10-14 days after
implantation, the limbs contained "muscles" made of Gelfoam, which
were not distinguishable in gross form and size from normal regenerates. At later stages, the Gelfoam regenerates became resorbed
in contrast to normal regenerates, the proximal portions of which
normally increase in diameter.

5.   Conclusions

At least on a superficial level, the main factors leading to
the morphogenesis of minced muscle regenerates have been worked out.
They are summarized below.

     a.   Gross morphogenesis can be accounted for on the basis of
(1) proximal and distal tendon connections which maintain the

elongated shape and (2) pressures of the surrounding tissues which
mold the overall shape. Work still needs to be done on the factors
which cause the proximal end of a muscle regenerate (or a normal
muscle) to be fleshy and the distal end to be thin and tendinous.

b. Tension appears to be the main force controlling the
establishment of a regular internal architecture. Tension is mediated by the connective tissue stroma, and the muscle fibers appear
to follow passively the configuration of the substrate.

c. In general although the control of cytodifferentiation
appears to be largely inherent in the myogenic cells themselves,
gross morphogenesis and the establishment of internal architecture
can be accounted for almost entirely on the basis of factors other
than the regenerating muscle fibers. These factors are primarily
physical interactions with other components of the limb.

II. LIMB REGENERATION

In dealing with the restoration of muscles in a regenerating
limb, one must always keep in mind that the muscular component is
inextricably linked with the overall epimorphic regenerative process.
In an amputated but nonregenerating limb the muscular component
responds to the trauma by undergoing a local regenerative process
which differs little in morphology and rate from the repair of a
transected or a minced muscle [5,7]. It is extremely difficult to
determine whether a given experimental manipulation is acting upon
a regenerating muscle itself or whether it is acting primarily upon
some other component of the regenerating limb and is only secondarily affecting the muscular component.

A. Initiation of Regeneration

The obvious initiating event in limb regeneration is amputation.
The mechanism by which amputation is translated into a stimulus for
regeneration has not been accurately determined. There is little
evidence for a discrete chemical stimulus which is released at the

time of amputation. Rather, amputation sets in motion a set of
tissue interactions which appear to sweep the tissues of the limb
stump into an overall regenerative response (reviewed in [13]). It
must not be forgotten that amputation is not the only way to elicit
an epimorphic regenerative process. Over the years many means have
been devised to stimulate the formation of supernumerary limbs.
This is also an epimorphic process, the fundamental features of
which do not differ qualitatively from those of post-amputational
limb regeneration [3]. Many of the ways of stimulating supernumerary limb formation, especially the use of foreign tissue implants,
involve the muscular component very early in the stimulatory process.

B. Reactions of Muscle during the Degenerative Phase

During the first few days following amputation, most of the
mesodermal tissues of the limb stump enter a phase of demolition.
The terminal ends of the transected muscle fibers undergo a local
degeneration, with a loss of cross striations and some sarcoplasmic
fragmentation. Macrophages play a role by removing damaged sarcoplasm, often leaving behind the empty basement membranes of the
muscle fibers [36]. The pH of the terminal segment of the limb
stump decreases, and the histochemical reactions of the degenerating
muscle fibers (summarized in [51]) are very similar to those which
occur in minced muscle regeneration. The morphology of both the
muscle and the entire limb stump during the degenerative phase is
so similar when regenerating and non-regenerating limbs are compared
that one cannot normally predict from histological sections whether
the limb is capable of regenerating or not.

C. Reactions of Muscle in the Dedifferentiative Phase

About a week or two after the amputation of an adult salamander
limb, there occurs an event which is the first good indication that
regeneration of the limb will occur. It is best illustrated in the
muscular tissues and is commonly called the phase of dedifferentiation. Histologically, dedifferentiation is represented by an

exaggerated loss of structure by many of the tissues underlying the
wound epidermis. In muscle the nuclei enlarge and the amount of
sarcoplasm diminishes before the tissue apparently breaks up.

There is still considerable debate concerning the cellular
nature of dedifferentiation, but most investigators [21,22,34,70]
feel that dedifferentiation in muscle is accomplished by the break-
ing off of mononucleated fragments from the damaged muscle fibers.
Such fragments are then said to lose their specialized cytological
characteristics and to transform into roughly spherical cells with
the following structural characteristics: increased nucleocyto-
plasmic ratio, one or two prominent nucleoli, relatively few mitochondria of small size, moderate to large amounts of rough endoplasmic reticulum and numerous Golgi complexes [23,34,48]. Later
these cells migrate distally to form the regeneration blastema.
Some investigators [35,50,51,73] doubt the existence of the dedifferentiative process as described above and consider fibroblasts or
other types of mononucleated cells as likely precursors of blastemal
cells. Too little is yet known about satellite cells in amphibian
muscle to assess properly their role, if any, in the epimorphic
regeneration of muscle. Although the cellular picture of events
during the dedifferentiative phase is still somewhat cloudy, the
characteristic histological picture of dedifferentiation heralds
the tentative commitment of the limb to regenerate. In the absence
of this process limb regeneration does not occur. Dedifferentiation,
however, is not an isolated event, but it is dependent upon a system
of tissue interactions in the limb stump. The broad details of this
system have been sketched out. The first essential component is the
presence of a wound epidermis covering the amputation surface. In
the early phases of regeneration the wound epidermis is underlain
by neither a basement membrane nor by a dermal layer, thus affording
direct contact between the epidermis and underlying mesodermal
tissues. In the absence of this contact, regeneration is prevented
[16,72], and little dedifferentiation occurs [Carlson, unpublished].

In normal limbs, however, a wound epidermis alone does not
suffice to bring about dedifferentiation. An adequate nerve supply
is also necessary. Normally, sprouts from the severed nerves invade
the wound epidermis [52,68], but other experiments have shown that
an approximation of nerves to the wound epidermis in lieu of actual
penetration of the epidermis is sufficient [55]. Singer [53,54]
has convincingly demonstrated that the requirement for nerves is
not a qualitative one with respect to the type of nerve, but it is,
instead, a quantitative one. In the absence of nerves, neither
muscle nor other tissues in the adult limb dedifferentiate to any
significant degree [45].

The wound epidermis and associated nerve supply must be accompanied by relatively recent mesodermal trauma in order for regeneration to occur. This, of course, is the rule in normal post-
amputational regeneration. The requirement for mesodermal trauma
was underscored by Polezhaev [41], who amputated the limbs of
axolotls and prevented regeneration by covering the amputation
surface with full thickness skin flaps. After two months he removed
the skin flaps from the ends of the limbs and allowed the former
amputation surfaces to re-epithelialize without causing trauma to
the underlying tissues. Regeneration did not occur. This observation has been confirmed in my laboratory [Carlson, unpublished].

In the case of supernumerary limb formation stimulated by the
subcutaneous insertion of tissue implants (e.g., frog kidney or
lung), some degeneration product of the implanted tissue seems able
to substitute for mesodermal trauma [14]. It is not known to what
extent the implant alone would suffice to bring about a dedifferentiative response without other tissue interactions.

In the absence of the conditions mentioned above, no regeneration of the limb occurs. The bone heals over with a cap of
cartilage and the soft tissues undergo limited regeneration and
heal over the end of the bone as a thin sheet of fibromuscular
tissue. The extensive loss of structure associated with dedifferentiation does not occur. A major analytical problem in the stage

of dedifferentiation is that with the involvement of so many components of the limb stump, it is very difficult to separate processes
specifically relating to the regeneration of muscles as discrete
entities from those processes involving regeneration of the entire
limb.

D. The Regeneration Blastema

After unspecialized mononuclear cells appear in the areas of
dedifferentiation (muscle, skeletal and connective tissues) of the
distal limb stump, they migrate distally to form a homogeneous-
appearing regeneration blastema beneath the wound epidermis, which
is thickened into a cap-like structure. The experiments of Thornton
[71] have shown that the thickened apical cap of epidermis attracts
the blastemal cells to their ultimate position beneath it. There
is again a difference of opinion concerning the role of the cells
derived from muscle with respect to their participation in the
regeneration blastema. Most investigators believe that muscular
elements form an integral part of the regeneration blastema.
However, the opinion has been expressed that the blastema is primarily derived from and will form only the skeleton and associated
connective tissue and that new muscle regenerates directly from the
stump musculature without participating in the blastema itself [26].
Whatever the case may be, during the period of the early blastema
it is not possible to identify myogenic cells in the regenerating
limb on the basis of their morphology alone. The blastema seems to
represent the key to the control of the regenerating limb. Whether
myogenic cells enter the blastema and receive morphogenetic in-
structions or whether their development is secondarily dependent
upon structures which arise from the blastema, the fact remains
that without the blastema further regeneration of muscles would not
occur.

E. Growth of the Limb Regenerate

The blastema next undergoes a phase of very rapid cell proliferation and elongation, during which the cartilaginous primordia of

the skeleton are laid down. Shortly thereafter the spindle-shaped
myoblastic cells aggregate in clumps alongside the skeletal elements.
Although opinion is not uniform, it is likely that some of the
muscle fibers regenerate as extensions of the stump musculature and
others differentiate distally in the absence of continuity with a
more proximal mass of muscle [29]. The differentiation of individual
muscle fibers does not appear to differ substantially from that of
other developing systems except that the early post-fusion state is
not so spectacularly delineated by regular centrally located nuclear
chains as in mammalian or even frog muscle.

F. Factors Controlling the Morphogenesis of Muscles in Limb Regenerates

Very little research has been directed toward factors controlling the morphogenesis of muscles themselves within regenerating
limbs. The most work, so far, has concerned the effect of foreign
muscle implants upon the overall morphogenesis of regenerating
limbs (reviewed by [17,75,76]). Aside from several references to
the control of form by ill-defined morphogenetic field forces, the
major recent experimental work and ideas have been those of Pietsch
[37-40]. As a result of transplanting regenerating limb blastemas
of larval Ambystoma into the caudal fin or the orbit with or without
a disk of underlying limb stump tissues, he concluded that morphogenesis of muscle in the regenerating limb is controlled by factors
in the stump or in the stump musculature. With this background in
the literature, I have recently begun an analysis of muscle morphogenesis in the regenerating limb.

The first experiment was designed to determine the effect of
lack of function on limb regeneration, specifically the muscular
component. The forelimbs of Ambystoma larvae were amputated, and
the animals were kept under continuous anesthesia (in 1:10,000 MS
222, Sandoz) starting either at the time of amputation or at early
phases of blastema formation. It was possible to keep these larvae
anesthetized continuously for periods of up to 16 days. Although

the limbs in anesthetized animals grew at a somewhat slower rate
than normal, the final structure of both the regenerated limbs as a
whole and the muscles within did not differ significantly from
normal controls [12].

The next parameter to be tested was the role of the stump
musculature in directing morphogenesis of muscles in limb regenerates. This problem was attacked directly by surgically removing
from 95-99% of the muscle from the upper arms of axolotls (130-
150 mm) and amputating the arms at the elbow. According to a
previous report by Holtzer [27], removal of the muscle from one
side of the tail in larval urodeles resulted in the virtual absence
of muscle in the corresponding side of the regenerate. In our
experiments [11] the limbs regenerated somewhat more slowly than
usual, but gross morphogenesis was perfect. Internally the distal
musculature usually approached the normal condition with respect to
overall form. In some cases almost every normal muscle could be
identified in the distal forelimbs of these regenerates. Although
this experiment offers little help in determining the origin of the
regenerating muscle fibers, it is clear that an intact stump musculature is not required for normal morphogenesis within a regenerating limb. This concept is supported by an experiment of Polezhaev
[42] in which he removed the underlying soft tissues from the limb
of an axolotl, chopped them into small pieces and stuffed them into
a casing composed of the normal skin of the limb stump. Although
he did not describe the morphology of the muscles in any detail, he
did illustrate a grossly normal limb regenerate containing a normal
skeleton. The apparently conflicting statements of Pietsch [37,38]
regarding morphogenetic direction by the muscles of the limb stump
can probably be reconciled by an examination of his experimental
model. Pietsch [37] transplanted young limb blastemas to the tail
fin of larvae with and without a disk of underlying stump tissue.
Under these experimental conditions it is reasonable to conclude
that the presence of the limb stump is necessary for normal muscles
to appear in a regenerate, but the role of stump musculature alone

was not tested. In orbitally transplanted limbs Pietsch [38] found
that if the orbital muscles were allowed to grow into the stump and
the limb was subsequently amputated, the pattern of the proximal
musculature was abnormal. This experiment indicates that the stump
musculature can influence the morphogenesis of regenerating muscles,
but how far apically the influence extends is not certain from the
information presented by Pietsch. The muscle extirpation experiments performed in my laboratory indicate that although the muscles
of the stump may normally exert some influence upon the form of the
muscles in the regenerate, their presence is not obligatory.

Beyond these few attempts to understand muscle morphogenesis
in the regenerating limb almost nothing is known. It is becoming
quite apparent that considerably more careful descriptive work must
be done in order to achieve a morphological basis for interpreting
correctly the results of experimental studies. Many important
questions remain to be answered. Among the more troubling are the
following:

1. Do the myogenic cells inherently contain or are they at
some time provided with any information which would enable them to
aggregate in a parallel fashion into a gross model of a muscle? Or
do they passively respond to other factors as seems to be the case
in minced muscle regeneration?

2. Is there any causal relationship between the regenerating
skeleton and the morphogenesis of the muscles which shortly appear
alongside? The studies of Pietsch [39] would indicate that there
isn't, but the disturbed morphology of muscles in limbs containing
skeletal defects suggests that there might be some influence of
skeleton upon muscle.

3. How much tension is exerted by the elongating skeletal
elements in a limb regenerate and is this a factor in determining
muscle morphogenesis?

4. What is the minimum number of components of the limb stump
necessary for normal morphogenesis of muscle? In the absence of

one component can another provide similar information to compensate
for that of the normal source?

5. How does the regeneration blastema play a role, if any, in
controlling the regeneration of muscles? The important question of
whether cells derived from muscle actually become part of the
blastema proper will have to be answered before any meaningful
experimental approach to this question can be designed.

6. Is the regeneration of muscles in limb regeneration an
integrated (and inseparable) part of the overall limb regenerative
process or can it be classified as a sub-process, proceeding according to its own ground rules?

III. CONCLUSIONS

Although quite a few apparent interactions between regenerating
muscles and other cells and tissues have come to light, our compre-
hension of the mechanisms of any of these is slight, at best. In
many instances gaps in our knowledge of the normal morphological
events during regeneration prevent anything but a superficial understanding of tissue interactions and control mechanisms. In many
cases the distinction between an interactive association of tissues
as opposed to an anatomical association must be kept in mind.

Any meaningful understanding of the stimulation of muscle
regeneration, whether of the tissue or epimorphic variety, must
await a definitive solution to the origin of the myogenic cell.
Until then it is only possible to relate a given stimulatory procedure to the subsequent appearance of cells which can be morphologically identified as myoblasts. In the case of minced muscle
regeneration it is presently impossible to decide which of the many
disturbances (trauma, denervation, ischemia, etc.) associated with
the operation is responsible for initiating the process. It is not
unlikely that each of these disturbances could serve as an effective
initial stimulus, but how this is transmitted to the potentially

myogenic cell is not known. In the case of limb regeneration, we
are confronted with the same dilemma.

In minced muscle regeneration, it is the return of the vascular
supply which seems to permit or to promote the progressive or differentiative phase of regeneration. Likely factors are a pronounced
shift in the metabolic environment of the revascularized areas
and/or the freeing of myogenic cells from the inhibiting influences
of the old sarcoplasm as a result of phagocytosis by blood borne
macophages. During limb regeneration, one does not see a simple
process of differentiation which shortly follows the initial
regenerative stimulus and the destructive phase. Rather, the
entire mesodermal component of the distal end of the limb stump
appears to be swept up into a rather exaggerated phase of loss of
adult structure (dedifferentiation) and the eventual aggregation of
the cells resulting from this process into the regeneration blastema.
During this time many cells and tissues of the limb must work in
concert with one another or the entire process is stopped. The
general pattern of differentiation of the regenerating limb is very
similar to that which occurs in the embryo.

From my vantage point [7], the regeneration of a minced muscle
represents a "hyperextension" of the healing capabilities of locally
damaged muscle. Provided a suitable metabolic environment, the
early myogenic elements seem able to begin their differentiation
with a minimal number of tissue interactions although later differentiation requires at least the influence of nerves. Morphogenesis seems to be determined largely by physical and mechanical
factors relating to the functional situation in which the regenerative process takes place, and it serves to illustrate the great
form regulatory capacities of the normally functioning limb.

In limb regeneration, on the other hand, the blastema appears
to represent a key to the subsequent regenerative events. During
the period of blastema formation significant redifferentiation of

injured muscle does not occur despite the fact that in minced
amphibian muscle full differentiation of striated muscle fibers has
occurred by the time a blastema has been established in an amputated
limb [5]. Exactly what is happening to the myogenic cells during
this period is entirely problematical. According to some viewpoints
they might be receiving morphogenetic information from either the
stump tissues or the blastema and according to others, they might
be remaining essentially in situ, from which location they would
grow distad and differentiate without acquiring any intrinsic
properties different from those of myoblasts in regenerating transected or minced muscle. In general, actual differentiation and
morphogenesis of new muscle tends to follow a proximo-distal sequence,
but this is not strictly true. Although existing data are not
sufficient to make any firm conclusions, it is not unlikely that the
proximal musculature of a limb regenerates from, and is strongly
influenced by, the normal stump musculature. Beyond this sphere of
influence, the morphogenesis of muscle appears to be an integral
part of the entire limb regenerative process. Whether a regenerating
muscle follows passively along pathways laid down by other components
of the developing limb or whether it possesses an intrinsic capacity
to guide to some extent its own development remains to be determined.

In not only the muscular component, but also the skeleton of a
regenerating limb, it seems that the most proximally damaged tissues
of the stump repair themselves by the "tissue" mode of regeneration
and that the differentiating muscular and skeletal elements follow
a different time course and obey different morphogenetic laws from
these same components of the more distal limb regenerate. Farther
apically, the bulk of the components of the regenerating limb appear
to be caught up in a complex system of developing cells and tissues
which are, of necessity, dependent upon interactions among one
another for the production of the perfectly formed regenerate.

REFERENCES

[1]   Allbrook, D. B., "An electron microscopic study of regenerating skeletal muscle," J. Anat., 96, 137-152 (1962).

[2]   Allbrook, D. B., W. de C. Baker, and W. H. Kirkaldy-Willis, "Muscle regeneration in experimental animals and in man," J. Bone and Joint Surg., 48B, 153-169 (1966).

[3]   Carlson, B. M., "Studies on the mechanism of implant-induced supernumerary limb formation in urodeles," I. The histology of supernumerary limb formation in the adult newt, Triturus viridescens," J. Exptl. Zool., 164, 227-242 (1967).

[4]   Carlson, B. M., "Regeneration of the completely excised gastrocnemius muscle in the frog and rat from minced muscle fragments," J. Morph., 125, 447-472 (1968).

[5]   Carlson, B. M., "The regeneration of a limb muscle in the axolotl from minced fragments," Anat. Rec., 166, 423-436 (1970).

[6]   Carlson, B. M., "Regeneration of the rat gastrocnemius muscle from sibling and non-sibling muscle fragments," Am. J. Anat., 128, 21-32 (1970).

[7]   Carlson, B. M., "Relationship between the tissue and epimorphic regeneration of muscles," Am. Zool., 10, 175-186 (1970).

[8]   Carlson, B. M., "The role of function in minced muscle regeneration" (abstr.), Anat. Rec., 166, 288 (1970).

[9]   Carlson, B. M., "Histological Observations on the Regeneration of Mammalian and Amphibian Striated Muscle," in Regeneration of Striated Muscle, and Myogenesis (A. Mauro, S. A. Shafiq and A. T. Milhorat, eds.) Exerpta Medica, Amsterdam, 1970, pp. 38-72.

[10]   Carlson, B. M., "The Regeneration of Entire Muscles from Minced Fragments," in Regeneration of Striated Muscle, and Myogenesis, (A. Mauro, S. A. Shafiq and A. T. Milhorat, eds.) Exerpta Medica, Amsterdam, 1970, pp. 25-37.

[11]   Carlson, B. M., "Muscle morphogenesis in limb regenerates following removal of stump musculature" (abstr.), Anat. Rec., 169, 289 (1971).

[12]   Carlson, B. M., "Organizational Aspects of Muscle Regeneration," in Research in Muscle Development and the Muscle Spindle, (R. J. Przybylski and B. Q. Banker, eds.) Exerpta Medica, Amsterdam, in press.

[13]   Carlson, B. M., "Factors Controlling the Initiation and Cessation of Early Events in the Regenerative Process," in Neoplasia and Cell Differentiation (G. V. Sherbet, ed.) Academic Press, New York, in press.

[14]   Carlson, B. M., "The distribution of supernumerary limb-inducing capacity in tissues of Rana pipiens," Oncology, 25, 365-371 (1971).

[15]   Gallucci, V., F. Novello, A. Margreth, and M. Aloisi, "Biochemical correlates of discontinuous muscle regeneration in the rat," Brit. J. Exptl. Pathol., 47, 215-227 (1966).

[16]   Godlewski, E., "Untersuchungen über Auslösung und Hemmung der Regeneration beim Axolotl," Arch. Entw-mech., 114, 108-143 (1928).

[17]   Goss, R. J., "Regeneration of vertebrate appendages," Advan. Morphogen., 1, 103-152 (1961).

[18]   Guth, L., "Trophic influences of nerve on muscle," Physiol. Revs., 48, 645-687 (1968).

[19]   Guth, L., "Trophic effects of vertebrate neurons," Neurosciences Res. Bull., 7, 1-73 (1969).

[20]   Gutmann, E. (editor), The Denervated Muscle, Publ. House Czechoslovak Acad. Sci., Prague, 1962.

[21]   Hay, E. D., "Electron microscopic observations of muscle differentiation in regenerating Amblystoma limbs," Devel. Biol., 1, 555-585 (1959).

[22]   Hay, E. D., "Cytological Studies of Dedifferentiation and Differentiation in Regenerating Amphibian Limbs." in Regeneration

(D. Rudnick, ed.) Ronald Press, New York, 1962, pp. 177-210.

[23]   Hay, E. D., Regeneration, Holt Rinehart Winston, New York, 1966.

[24]   Hess, A. and S. Rosner, "The satellite cell bud and myoblast in denervated mammalian muscle fibers," Am. J. Anat., 129, 21-40 (1970).

[25]   Hikida, R. S. and W. J. Bock, "The structure of pigeon muscle and its changes due to tenotomy," J. Exp. Zool., 175, 343-356 (1970).

[26]   Holtzer, H., in discussion of T. P. Steen, "Cell differentiation during salamander limb regeneration," in Regeneration of Striated Muscle, and Myogenesis (A. Mauro, S. A. Shafiq and A. T. Milhorat, eds.) Exerpta Medica, Amsterdam, 1970, pp. 73-90.

[27]   Holtzer, S., "The inductive activity of the spinal cord in urodele tail regeneration," J. Morphol., 99, 1-39 (1956).

[28]   Hsu, L., "The effect of innervation on minced muscle regeneration in frogs" (abstr.), Anat. Rec., 166, 321 (1970).

[29]   Laufer, H., "Immunochemical studies of muscle proteins in mature and regenerating limbs of the adult newt, Triturus viridescens," J. Embryol. Exptl. Morph., 7, 431-458 (1959).

[30]   Lee, J. C., "Electron microscopic observations on myogenic free cells of denervated skeletal muscle," Exptl. Neurol., 12, 123-135 (1965).

[31]   Le Gros Clark, W. E., "An experimental study of the regeneration of mammalian striped muscle," J. Anat., 80, 24-36 (1964).

[32]   Le Gros Clark, W. E. and L. B. Blomfield, "The efficiency of intramuscular anastomoses, with observations on the regeneration of devascularized muscle," J. Anat., 79, 15-32 (1945).

[33]   Le Gros Clark, W. E. and H. S. Wajda, "The growth and maturation of regenerating striated muscle," J. Anat., 81, 56-63 (1947).

[34]   Lentz, T. L., "Cytological studies of muscle dedifferentiation

and differentiation during limb regeneration of the newt Triturus,"
Am. J. Anat., 124, 447-480 (1969).

[35] Manner, H. W., "The origin of the blastema and of new tissues
in regenerating forelimbs of adult Triturus viridescens
(Rafinesque)," J. Exptl. Zool., 122, 222-257 (1953).

[36] Norman, W. P. and A. J. Schmidt, "The fine structure of
tissues in the amputated regenerating limb of the adult newt,
Diemictylus viridescens," J. Morphol., 123, 271-312 (1967).

[37] Pietsch, P., "Differentiation in regeneration," I. The
development of muscle and cartilage following deplantation of regenerating limb blastemata of Amblystoma larvae, Devel. Biol.,
3, 255-264 (1961).

[38] Pietsch, P., "The effects of heterotopic musculature on myogenesis during limb regeneration in Amblystoma larvae," Anat. Rec.,
141, 295-303 (1961).

[39] Pietsch, P., "Independence of chondrogenesis from myogenesis
during limb regeneration in Amblystoma larvae," J. Exptl. Zool.,
150, 119-127 (1962).

[40] Pietsch, P., "The influence of spinal cord on differentiation
of skeletal muscle in regenerating limb blastema of Amblystoma
larvae," Anat. Rec., 142, 169-178 (1962).

[41] Polezhaev, L. V., "Concerning processes of resorption, proliferation and relationships of tissues during regeneration of
limbs in axolotls" (Russian), Biol Zhur., 2, 368-386 (1933).

[42] Polezhaev, L. V., "Concerning the Determination of Regenerates"
(Russian), in To Academician N. V. Nassonov (Russian), Izdatel.
Akad. Nauk SSSR, Moscow, 1937, pp. 151-247.

[43] Popova, M. F., "Some biochemical properties of regenerating
skeletal muscle" (Russian), Arch. Anat. Gist. Embryol., 39, 60-64
(1960).

[44] Reznik, M., "Origin of myoblasts during skeletal muscle

regeneration," Lab. Investig., 20, 353-363 (1969).

[45]   Rose, S. M., "The role of nerves in amphibian limb regeneration," Ann. N. Y. Acad. Sci., 49, 818-833 (1948).

[46]   Rumyantseva, O. N., "Development of minced muscle tissue
transplanted under the skin" (Russian), Doklady Akad. Nauk SSSR.,
125, 435-438 (1959).

[47]   Rumyantseva, O. N., "New findings on the role of tension in
the differentiation of myogenous tissue" (Russian), Arkh. Anat.
Gist. Embryol., 39, 51-59 (1960).

[48]   Salpeter, M. M. and M. Singer, "The fine structure of mesen-
chymatous cells in the regenerating forelimb of the adult newt,
Triturus," Devel. Biol., 2, 516-534 (1960).

[49]   Samsonenko, R. V., "Development of muscle tissue following
the transplantation of minced muscle in place of completely removed
muscles in the frog" (Russian), Arkh. Anat. Gist. Embryol., 33,
56-64 (1956).

[50]   Schmidt, A. J., "Distribution of polysaccharides in the regenerating forelimb of the adult newt, Diemictylus viridescens,"
J. Exptl. Zool., 149, 171-192 (1962).

[51]   Schmidt, A. J., Cellular Biology of Vertebrate Regeneration
and Repair, Univ. Chicago Press, Chicago, 1968.

[52]   Singer, M., "The invasion of the epidermis of the regenerating
forelimb of the urodele, Triturus, by nerve fibers," J. Exptl. Zool.,
111, 189-210 (1949).

[53]   Singer, M., "The influence of the nerve in regeneration of
the amphibian extremity," Quart. Rev. Biol., 27, 169-200 (1952).

[54]   Singer, M., "Induction of regeneration of the forelimb of the
postmetamorphic frog by augmentation of the nerve supply," J. Exptl.
Zool., 126, 419-471 (1954).

[55]   Singer, M. and S. Inoue, "The nerve and the epidermal cap in
regeneration of the forelimb of adult Triturus," J. Exptl. Zool.,
155, 105-116 (1964).

[56] Snow, M. H., "A histochemical study of minced muscle regeneration in the rat" (abstr.), Anat. Rec., 166, 381 (1970).

[57] Snow, M. H., "Glycolysis in degenerating skeletal muscle isolated from early minced muscle regenerates in the rat" (abstr.), Anat. Rec., 169, 433 (1971).

[58] Striganova, A. R., Reactivity and Restorative Capacity of Denervated Muscle at Different Stages of Atrophy (Russian), Izdatel. Akad. Nauk SSSR, Moscow, 1961.

[59] Studitsky, A. N., "Restoration of muscle by means of transplantation of minced muscle tissue" (Russian), Doklady Akad. Nauk SSSR., 84, 389-392 (1952).

[60] Studitsky, A. N., "Principles of the restoration of muscle in higher vertebrates" (Russian), Trudy Inst. Morph. Zhivotnikh im A. N. Severtsova., 11, 225-264 (1954).

[61] Studitsky, A. N., Experimental Surgery of Muscles (Russian), Izdatel. Akad. Nauk SSSR, Moscow, 1959.

[62] Studitsky, A. N., "Dynamics of the Development of Myogenic Tissue under Conditions of Explantation and Transplantation," in Cinemicrography in Cell Biology (G. G. Rose, ed.) Academic, New York, 1963, pp. 171-200.

[63] Studitsky, A. N., "Free auto- and homografts of muscle tissue in experiments on animals," Ann. N. Y. Acad. Sci., 120, 789-801 (1964).

[64] Studitsky, A. N. and Z. P. Ignatieva, The Restoration of Muscle in Higher Mammals (Russian), Izdatel. Akad. Nauk SSSR, Moscow, 1961.

[65] Studitsky, A. N. and R. P. Zhenevskaya, Theory and Practice of the Auto- and Homotransplantation of Muscles, Publishing House Nauka, Moscow, 1967.

[66] Studitsky, A. N., R. P. Zhenevskaya, and O. N. Rumyantseva, "Fundamentals of the technique of restoration of muscle by means of

transplantation of minced muscle tissue" (Russian), Cesk. Morfol.,
4, 331-340 (1956).

[67] Studitsky, A. N., R. Zhenevskaya, and O. Rumyantseva, "The
Role of Neurotrophic Influences upon the Restitution of Structure
and Function of Regenerating Muscle," in The Effect of Use and
Disuse on Neuromuscular Functions (E. Gutmann and P. Hník, eds.),
Publ. House of Czechoslovak Acad. Sci., Prague, 1963, pp. 71-81.

[68] Taban, C., "Les fibres nerveuses et l'epithelium dan l'edi-
fication des régénérates de pattes (in situ ou induites) chez le
Triton," Arch. Sci., 2, 553-561 (1949).

[69] Taban, C., "Quelques problemes de régénération chez les
urodèles," Rev. Suisse Zool., 62, 387-468 (1955).

[70] Thornton, C. S., "The histogenesis of muscle in the regenerating fore limb of larval Amblystoma punctatum," J. Morphol., 17-47
(1938).

[71] Thornton, C. S., "Influence of the Wound Skin on Blastemal
Cell Aggregation," in Regeneration in Animals and Related Problems
(V. Kiortsis and H. A. L. Tramphsch, eds.), North-Holland Publ. Co.,
Amsterdam, 1965, pp. 333-339.

[72] Tornier, G., "Kampf der Gewebe im Regeneration bei Begünstigung
der Hautregeneration," Arch. Entw-mech., 22, 348-369 (1906).

[73] Toto, P. D. and J. D. Annoni, "Histogenesis of newt blastema,"
J. Dental Res., 44, 71-79 (1965).

[74] Trupin, G. L., "An ultrastructural study of the regeneration
of the minced gastrocnemius muscle in the frog" (abstr.), Anat. Rec.,
166, 391 (1970).

[75] Vorontsova, M. A., Regeneration of Organs in Animals (Russian),
Izdatel. Sovietskaya Nauka, Moscow, 1949.

[76] Vorontsova, M. A. and L. D. Liosner, Asexual Propagation and
Regeneration, Pergamon Press, New York, 1960.

[77] Yeasting, R. A., "The effect of the nerve supply on the regeneration of minced skeletal muscle in the mouse," Ph.D. Dissertation, University of Louisville, Louisville, Kentucky, 1969.

[78] Zhenevskaya, R. P., "The role played by nervous connections in the early stages of muscle regeneration" (Russian), Doklady Akad. Nauk SSSR., 121, 182-185 (1958).

[79] Zhenevskaya, R. P., "The influence of differentiation upon the regeneration of skeletal muscle" (Russian), Arkh. Anat. Gist. Embryol., 39, 42-50 (1960).

[80] Zhenevskaya, R. P., "Restoration of muscle by the method of transplantation of minced muscle tissue under conditions of sensory denervation" (Russian), Arkh. Anat. Gist. Embryol., 40, 46-53 (1961).

[81] Zhenevskaya, R. P., "Experimental histologic investigation of striated muscle tissue," Rev. Can. Biol., 21, 457-470 (1962).

CHAPTER 3

ENZYME KINETICS AND THE MECHANISM OF MUSCLE CONTRACTION

Edwin W. Taylor and Richard W. Lymn

Department of Biophysics
University of Chicago
Chicago, Illinois

There are few biological problems which have been subjected to such an extensive investigation as the mechanism of muscle contraction. In many respects the muscle provides a model system. It will function outside the animal and is sufficiently regular to be studied by X-ray diffraction at rest and in the excited state. The mechanical and thermal properties can be measured and energy liberation can be correlated with ATP hydrolysis. Finally, the protein components are readily obtained in large quantities for enzymatic and physical studies.

Although there may be some disagreement as to the finer details, a general picture has emerged, at least for the contraction of striated muscle, which might be referred to as the sliding filament-movable bridge model. Electron microscopic and X-ray diffraction studies by Huxley, Hanson, Pringle, Eliot, Lowy, Worthington, Reedy, Tregear, and others have led to essentially the following description. The contractile unit is the sarcomere, which consists of two interdigitating sets of thick (myosin) and thin (actin-tropomyosin-troponin) filaments. Contraction consists of a sliding of the thin filaments into the thick filament lattice (the sliding filament model of Hanson and Huxley). There is no measurable shortening of individual thick or thin filaments. The only connections between

the two lattices and therefore the only means of exerting tension
are a set of regularly spaced "bridges" projecting from the thick
filaments. The bridges are the globular portions of myosin molecules and the ATP and actin binding sites are located in these
globular regions.

It could be argued that there are long range forces between
the two lattices which at least account for the lateral filament
spacings in resting muscle [4]. Such forces may also be important
during contraction [5] but in view of strong binding between actin
and myosin in solution, short range chemical and electrostatic
forces seem likely to predominate.

Evidence from X-ray diffraction does not prove that the movement of bridges causes contraction but the results are suggestive
and certainly consistent with this hypothesis. Namely, in relaxed
muscle the mass distribution is such that the ends of the bridges
are not in contact with actin filaments [11]. In rigor, the muscle
has a high resistance to stretch which is easily explained if the
bridges are attached to actin filaments. Again the X-ray evidence
indicates that mass has moved outward from the thick toward thin
filaments so that they are in contact through the cross bridges [11].

Diffraction patterns of excited muscle fail to show the re-
flections arising from the spacing between the bridges [11]. This
could be explained by a nonsynchronous movement of the bridges in
the excited state. They no longer diffract because the spacing
between them is now distributed over the range determined by the
extent of movement of the active bridge [Fig. 1 (a)]. The bridges
are arranged in pairs every 143 Å and give rise to a helical array
of six bridges in 429 Å. The movement of the end of the active
bridge could be small, i.e., a few angstroms, although it appears
more reasonable from the loss in scattering in the active state
that the end moves by a distance at least comparable to the thickness of the bridge, or about 50-100 Å. It is unlikely that the
movement could be larger than 100 Å since the globular region is

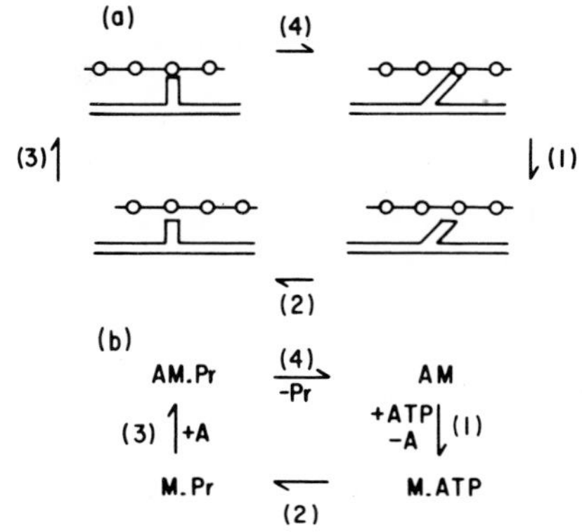

Figure 1.   (a) Contractile cycle for a single cross bridge.
Actin sites which are suitably oriented to interact with a cross
bridge are indicated by circles.   (b) Steps in chemical mechanism
for ATP hydrolysis by actomyosin.   Binding of ATP and dissociation
of actin is shown as a single step because actin dissociation is
very fast following substrate binding.   Pr stands for reaction
products, ADP and phosphate.

only 150–200 Å in length.   Further support for bridge movement comes
from electron microscopic studies of insect flight muscle.   In
rigor, bridges appear to be bent back at about a 45° angle with the
tip pointing away from the Z-line [15].   In negatively stained
preparations of actin filaments decorated with heavy meromyosin
molecules the heads again point out from the filament at 45° in the
direction away from the Z-line [9].

The evidence, though not entirely convincing, leads to a model
for the molecular contraction cycle [Fig. 1 (a)], drawn by a number
of authors [8,10,14].   There are certain features which should be
stressed.   The contraction cycle requires at least four steps.
Namely, dissociation of the bridge, movement of the free bridge,
recombination with the actin filament, and return to the original
orientation with translation by the unit step size which we take to

be 50-100 Å. It would be difficult to think of a moving bridge
model with a smaller number of distinguishable steps.

In numbering the steps in the sequence shown, the implication
is a drive stroke from the straight to the bent configuration,
based on the results for insect flight muscle. The diagram could
have been drawn with the direction of all the arrows reversed without greatly altering the arguments.

This qualitative model can be made into a working model from
which the relation of force to velocity and rate of energy liberation can be calculated. To do this one assumes values for the rate
constants of the steps and assigns to the bridge the mechanical
properties of a linear spring. A theory of this type was developed
by A. F. Huxley in 1957 [8]. Considering what was then known about
bridge movement and actomyosin chemistry, Huxley's theory showed
remarkable insight and it has continued to dominate model building
ever since.

How does enzyme chemistry contribute to solving the mechanism?
The answer is clear if we look back at the molecular cycle shown in
Fig. 1 (a). This cycle or the A. F. Huxley model is a kinetic
scheme. The steps in the cycle should correspond in some way to
the enzyme mechanism describing the hydrolysis of ATP by actomyosin.
Hopefully the enzyme mechanism should "make sense" when we try to
fit it together with the postulated contraction cycle. It should
in fact tell us what steps occur in the cycle and supply some of
the rate constants necessary for a quantitative model.

The simplest contraction cycle has three or four steps and
studies of ATP hydrolysis in the steady state give only the rate of
the slowest step. In order to observe a sequence of steps the most
direct approach is to study the transient state, preferably on a
time scale comparable to events taking place in the muscle.

We have pursued this approach to the problem in the past few
years and the purpose of the remainder of this paper is to present
a simple enzyme scheme for ATP hydrolysis by actomyosin and to
attempt to relate it to the contraction cycle.

There are a number of well known properties of actomyosin and myosin which are directly related to the muscle mechanism. Under physiological conditions (0.1-0.15 M KCl, greater than 1 mM $MgCl_2$) myosin is a very poor ATPase. The activity is much higher if Ca is substituted for Mg but this fact is probably not relevant since the Ca to Mg ratio in muscle will always be extremely small. In the absence of any divalent ions, that is to say, in the presence of a chelating agent such as EDTA, the rate of hydrolysis can be increased by a factor of 200-500 fold. The ionic strength has to be relatively high to obtain this activation, at least in part because the charge on ATP must be neutralized by binding K^+ if Mg^{2+} is absent. It would be more proper to describe myosin as an enzyme which is strongly inhibited by Mg^{2+} and weakly inhibited by Ca^{2+}. Inhibition of an ATPase by Mg^{2+} is an uncommon result and in our view it is an important property of the physiological mechanism.

The second property of myosin is the "early burst" effect first described by Weber and Hasselbach [19] and extensively studied by Tonomura [18]. The time course of hydrolysis when ATP is mixed with myosin under physiological conditions shows a rapid early phase during which one to two molecules of ATP are hydrolyzed per myosin. This is followed by a slow rate of further hydrolysis in the steady state. We have studied this early phase extensively by fast kinetic methods, either by monitoring the hydrogen ion which should be formed when ATP is hydrolyzed at pH 8 or by directly measuring phosphate formation in a rapid mixing apparatus [6,7,13, 17]. The time course of the early phosphate production in a typical experiment is shown in Fig. 2. The early phase proved to be extremely fast indicating that myosin was capable of hydrolyzing the first molecule of ATP per site at a rate of about 100 sec^{-1}. This conclusion was reached in the following way. The simplest mechanism for the reaction up to the step in which phosphate is formed is

$$M + ATP \underset{k-1}{\overset{k_1}{\rightleftharpoons}} M \cdot ATP \overset{k_2}{\rightleftharpoons} M \cdot ADP \cdot P$$

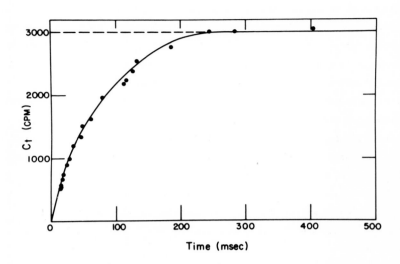

Figure 2. Phosphate production with MgATP as substrate.
0.5 M KCl, 0.1 M Tris, 0.01 M $MgCl_2$, pH 8.0, 20°C, 32 μM ATP,
2 mg/ml myosin. The dotted line is an extrapolation of the steady
state to zero time. The burst is 1.2 moles P_i/mole myosin.

The nature of the myosin product complex need not be specified; we
only wish to indicate the bond has been hydrolyzed in step 2. If
k_1[ATP] << k_2 the apparent rate should increase linearly with ATP
concentration. If k_1[ATP] $\overset{>}{-}$ k_2 the rate should reach a maximum
value. The rate constant at a particular ATP concentration was
determined from a semi-log plot of phosphate formation versus time
during the transient phase. The apparent rate constant was then
measured as a function of ATP concentration with the results shown
in Fig. 3. Although there is some scatter in the data it is reasonable to conclude that expectations of the simple mechanism are ful-
filled. The rate is proportional to ATP concentration and from the
slope we obtain the rate constant k_1 which is the order of 10^6
$M^{-1}sec^{-1}$ in 0.1 M KCl, pH 8, 20°C. The plateau value k_2 is approximately 100 sec^{-1} at physiological ionic strength. The important
point to be made is that myosin hydrolyzes ATP at a very rapid rate.

The value of k_2 is larger than the rate for myosin fully activated
by actin. The low activity of myosin in the steady state is due to
a slow step which follows hydrolysis. By measuring the rate of
decay of the product complex as myosin moves rapidly down a Sephadex
G-25 column, the rate of the slow step was determined. This rate
was found to be nearly equal to the steady state rate of hydrolysis.
Thus the next step is simply

$$M \cdot ADP \cdot P \xrightarrow{k_3} M + ADP + P$$

It has been suggested, particularly by Tonomura [18], that the early
burst, since it is stoichiometric, is evidence for a phosphorylated
intermediate. Our evidence, though not disproving this possibility,
makes it appear extremely unlikely. A stable intermediate has never

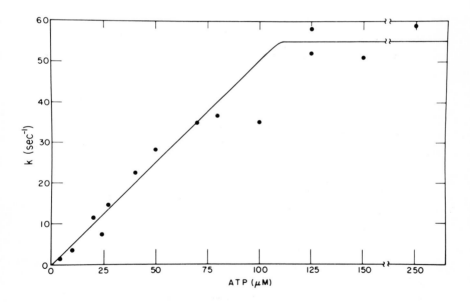

Figure 3. The variation of the apparent pre-steady state rate
constant of P_i production versus MgATP concentration. All solutions
0.5 M KCl, 0.1 M Tris, 0.01 M $MgCl_2$, pH 8.0, 20°C. Myosin 1.5-3
mg/ml.

been isolated. The column experiments show that both ADP and phosphate are bound and ADP dissociates at a slightly slower rate than
phosphate. All that is required to give a stoichiometric early burst
is that $k_3 \ll k_2$ and the measured values are 0.05 and 100 sec^{-1}. The
studies of Sartorelli et al. [16] also failed to demonstrate a
covalent intermediate by ^{18}O exchange. Although we cannot rule out
the possibility that the mechanism is more complex than we have
indicated in the simple scheme and the system may pass through a
covalent intermediate, there is no evidence for such an intermediate.
The results are all explainable if the rate limiting step is the
dissociation of ADP from the enzyme. ADP binding is enhanced by
Mg^{2+} which is presumably the cause of Mg inhibition and its reversal
by EDTA.

For the present our view is that there are at least three steps
in the mechanism and reasonable values can be assigned to the rate
constants. There may be other intermediates between the enzyme
substrate and enzyme-products complex but if so they have not yet
been characterized.

We now wish to consider the mechanism of actomyosin hydrolysis
of ATP and to describe some recent work on this problem. Actin and
myosin form a relatively strong complex yet when ATP is added the
system appears to dissociate based on measurements of turbidity,
viscosity, flow birefringence or sedimentation pattern. At low
ionic strength the behavior is complex, dissociation occurs under
some conditions and is then followed by precipitation and super-
precipitation. In the superprecipitated state the ATPase activity
is 10-20 times larger than for myosin alone. These last two properties have generally been regarded as the analogue in the dis-
ordered system, of contraction and activation of muscle. However a
heterogeneous system is not suitable for kinetic analysis. In any
case the degree of activation is so small compared to energy output
of resting versus active muscles (a factor of two or three thousand)
that any inferences from kinetics would be highly questionable.

A significant step was made by Eisenberg and Moos employing acto heavy meromyosin (acto-HMM) which does not superprecipitate [2,3]. They recognized that two factors were involved, the increase in enzyme activity and the dissociation of the system by ATP and that these are competing reactions. By measuring enzyme activity as a function of actin concentration and extrapolating to an infinite actin concentration the real activation factor could be determined. A value of about 200-fold was obtained.

The Eisenberg and Moos scheme can be expressed as

$$AM + ATP \rightleftharpoons AM \cdot ATP \xrightarrow{\text{fast}} AM + ADP + P$$
$$\pm A \updownarrow$$
$$M \cdot ATP \xrightarrow{\text{slow}} M + ADP + P$$

Dissociation by ATP is attributed to reduction in the association constant of AM · ATP compared with AM. This scheme accounts for the steady state kinetics and the dependence of the rate on actin concentration. However, the maximum rate of hydrolysis is still only about 10-20 $\sec^{-1}$ which is less than the rate at which myosin alone is capable of hydrolyzing the first molecule of ATP. The role of actin therefore need not be to enhance the rate of the actual hydrolysis step but rather to increase the rate of product dissociation. This at first appears to be a minor change and could be incorporated in the above scheme by simply adding an AM · ADP · P intermediate. But in addition we require the kinetic scheme to satisfactorily account for the transient state behavior and here we run into a difficulty.

The early burst phase was measured for acto-HMM and the behavior of the system was quite similar to HMM alone except for the increased steady state rate. Figure 4 shows early phosphate liberation for acto-HMM and HMM alone. The size of the early burst is unchanged and the rate during the early phase is somewhat reduced. When the rate was measured as a function of ATP concentration, results shown in Fig. 5 were obtained. The rates are rather difficult to measure because the reaction is very fast at low ionic strength. However, we chose to work at this low ionic strength to insure that there would be a large activation of the steady state rate by actin.

The main points about the behavior are the following. At a given substrate concentration the apparent rate constant decreased with increasing actin concentration up to about three G-actin per HMM which is approximately the ratio at which all HMM is bound to actin. At a high actin concentration exceeding 3-4 actin per HMM the increase in rate with ATP concentration was approximately proportional to ATP concentration. These results indicate that the rate constant for substrate binding to acto-HMM is less than for HMM. The actual values were roughly 1.5×10^6 $M^{-1}sec^{-1}$ and 2×10^6 $M^{-1}sec^{-1}$, respectively, at this ionic strength. At high ATP concentrations the plot shows appreciable curvature, but in the accessible concentration range the plateau was not reached.

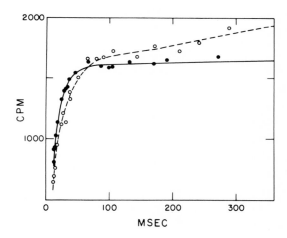

Figure 4. Early phase in the hydrolysis of ATP by HMM and acto-HMM. —•—•— phosphate liberation by HMM, —o—o— phosphate liberation by acto-HMM. Conditions 0.05 M KCl and 0.02 M Tris buffer, pH 8, 20°C, 5×10^{-3} M $MgCl_2$, 25 μM ATP, 4.8 moles actin/mole HMM, 1 mg/ml HMM. Intercept of the linear portion corresponds to an early burst of approximately 1.1 moles phosphate/mole HMM.

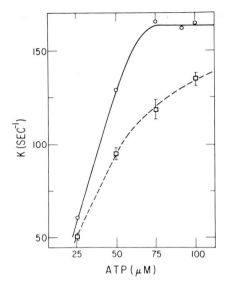

Figure 5. Pre-steady state rate of phosphate production by HMM
and acto-HMM, 0.05 M KCl, 5 x 10^{-3} $MgCl_2$, pH 8, 20°C. —o—o— HMM,
—Φ-Φ—acto-HMM. For each substrate concentration the rate constant
was determined for two or more actin to HMM ratios in the range from
0.4 to 2.16 actin to HMM by weight. There was no significant variation with actin to HMM ratio, and data was averaged for each substrate concentration. Standard deviations are indicated by the
error bars. The points are average values of 13 experiments.

Very roughly it appears that the maximum rate is not larger than
for HMM alone, and is the order of 100-125 sec^{-1}.

    Thus the kinetic behavior is very similar to that of HMM alone,
the main difference being a 10-20 fold increase in the steady state
rate.

    An important question now arises. Once the AM · ATP complex
is formed there are two alternatives for the next step

    AM · ATP ⟶ AM · ADP · P   or   AM · ATP ⟶ A + M · ATP

Which alternative is the correct one will be determined by which

rate constant is larger, the hydrolysis of the bond or the dissociation of the protein complex. In previous studies with actomyosin
at high ionic strength the rate of dissociation was measured by the
change in turbidity using a stop-flow apparatus. The rate of dissociation was strictly proportional to ATP concentration and showed
no sign of reaching a maximum rate. Similar experiments have been
done with the acto-HMM system with a similar result. The rate of
dissociation increases with ATP concentration and becomes too fast
to measure. By examining the turbidity change before flow stops a
very rough estimate of the rate is obtained by measuring the fraction of turbidity change relative to the maximum change in turbidity.
The time for the solution to travel from the mixer to the observation window is about 3 msec at the maximum flow rate. At a high
ATP concentration the turbidity change is complete in 3 msec which
means that the apparent rate of dissociation is larger than 700-1000
sec^{-1}. A series of traces for increasing ATP concentrations is
shown in Fig. 6.

We conclude that the system dissociates following AM · ATP
complex formation at a rate which is much faster than bond hydrolysis.
It has been known for a long time that hydrolysis is not necessary
for dissociation since pyrophosphate is capable of dissociating
actomyosin.

It therefore appears that in our experiments dissociation must
have preceded hydrolysis and the early burst occurred on the free
HMM · ATP. The experiments then did not measure hydrolysis for
AM · ATP and we do not know the rate constant for this step.

If AM dissociates before the bond is hydrolyzed how does actin
activate the enzyme? The most reasonable explanation is provided
by writing out the simplest kinetic scheme and noting that there is
an alternate pathway to the AM · ADP · P complex.

$$AM + ATP \xrightleftharpoons{(4)} AM \cdot ATP \qquad AM \cdot ADP \cdot P \xrightleftharpoons{(7)} AM + ADP + P$$

$$\pm A \updownarrow (5) \qquad\qquad \pm A \updownarrow (6)$$

$$M + ATP \xrightleftharpoons{(1)} M \cdot ATP \xrightleftharpoons{(2)} M \cdot ADP \cdot P \xrightarrow{(3)} M + ADP + P$$

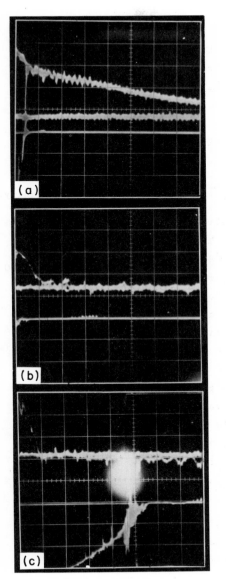

Figure 6. Dissociation of acto-HMM induced by ATP.

(a) 10^{-5} M ATP, 5×10^{-3} M MgCl$_2$, 0.1 M KCl, 0.05 Tris buffer, pH 8, HMM, 5×10^{-6} M, actin 2.5×10^{-5} M, 10°C; time scale 100 msec per major division. The lower trace is a signal monitoring the position of the drive block. Flow stops at about 50 msec after the beginning of the sweep. A second sweep a few seconds later gave the thick horizontal line indicating the final turbidity. The half life of the turbidity change is 450 msec.

(b) Conditions as in (a) except the ATP concentration was 5×10^{-4} M; time scale 10 msec per major division. Flow stops approximately 2 msec after the beginning of the sweep. The half life is approximately 7 msec.

(c) 10^{-2} M ATP, 1.5×10^{-2} M MgCl$_2$, 20°C, other conditions as in (a). Time scale 20 msec per major division. Sweep was triggered at the beginning of the drive. Rising trace indicates the position of the drive block. Flow stops near the middle of the trace. The turbidity falls to the final value long before the flow stops, indicating reaction is much too fast to measure (half life less than 1 msec). The light area at the center of the figure is a reflection from room lights.

For each numbered step in the scheme we assign a rate constant $k_{\pm i}$.
The mechanism is consistent with the transient state evidence as
well as the steady state results of Eisenberg and Moos [2,3]. The
steps in the mechanism are the displacement of actin from AM by the
binding of ATP, the ATP is hydrolyzed by free myosin to give
M · ADP · P which decays very slowly. Activation is brought about
by the binding of A to the M · ADP · P complex followed by rapid
dissociation of the products from the AM · ADP · P complex.

Since this mechanism appears to be rather novel we have tried
to demonstrate that step (6) does indeed occur. A simple two-stage
mixer was constructed to first mix M with H^3-ATP-γ^{32}P and then in
1-2 sec with actin plus unlabeled ATP. In this time interval any
M · ATP will have been converted to M · ADP · P. After mixing with
actin the solution was applied to a Sephadex column and rapidly
eluted. The results of typical experiments are shown in Fig. 7.
Each tube was assayed for ADP and phosphate in the experiment shown
in Fig. 7 (a). In the absence of actin there is a peak of ADP and
phosphate which elutes with the myosin and a trail of radioactivity
due to slow dissociation as myosin moves down the column. Mixing
with actin essentially displaces the products from the complex.
The small shoulder which remains is probably due to incomplete
mixing. This experiment does not measure the rate constants, it
simply shows that step (6) will take place at a reasonable rate.
A similar result was obtained using pyrophosphate, which binds
strongly to myosin and also causes dissociation of AM. Mixing with
actin quantitatively displaced pyrophosphate from the myosin
[Fig. 7 (c)].

The scheme we have presented is tentative and further studies
are still in progress. There is little doubt that it is oversimplified but it provides a simple kinetic model which can be
related to the contraction cycle. In Table 1 we have listed a preliminary set of rate constants for the scheme which may be useful
for those who wish to speculate. k_6 has not been directly measured
but can be inferred from the increase in steady-state rate with

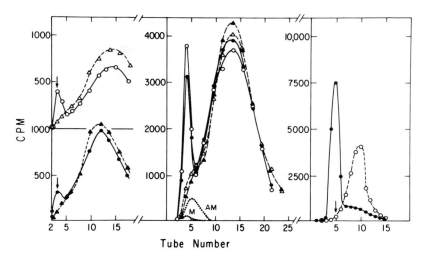

Figure 7. (a) Dissociation of myosin-product complex by actin. Conditions, 0.1 M KCl, 8°C, 0.05 M Tris buffer, pH 8, 5×10^{-3} M $MgCl_2$, 0.25 ml myosin 3.5×10^{-5} M mixed with 0.25 ml ^3H ^{32}P ATP (2×10^{-5} M) and further mixed after 1-2 sec with 1 ml actin (7×10^{-5} M) plus unlabeled ATP (10^{-3} M) (---) or ATP alone (——). Protein peak, indicated by arrow, eluted at approximately 40 sec after mixing. Column buffer contained 10^{-3} M ATP, ADP, and phosphate. Three-milliliter fractions were collected in tubes containing PCA and carrier ATP, ADP, and P and each tube was assayed for ADP and phosphate. —•—•— phosphate, no actin —o—o— ADP, no actin —▲—▲— phosphate, actin present —△—△—, actin present. (b) Dissociation of myosin-product complex by actin. Conditions 0.3 M KCl, 7°C, 0.05 M Tris buffer, pH 8, 5×10^{-3} M $MgCl_2$. 0.25 ml myosin (1.8×10^{-5} M) mixed with 0.25 ml ^3H ^{32}P ATP (2×10^{-5} M) and further mixed with 1 ml of 6×10^{-5} M actin plus 10^{-3} ATP (---) or ATP alone (——); 3 ml per tube, protein peak eluted at 45 sec (no actin) and 50 sec (plus actin). —•—•— ^{32}P, no actin —o—o— ^3H, no actin —▲—▲— ^{32}P, actin present —△—△— ^3H, actin present. Positions of protein peaks determined by absorption at 276 nm are given by dotted curves; M and AM refer to myosin and actomyosin, respectively. (c) Dissociation of myosin-pyrophosphate complex by actin. HMM (10^{-5} M)

actin concentration. k_7 was estimated from data of Eisenberg and
Moos [2,3], since it must be the rate limiting step at high ATP and
actin concentrations. As is discussed in the next section, the
magnitude of k_7 is crucial for deciding among models of contraction
and it is important to find a way to measure this rate directly
rather than to infer it by an extrapolation.

The kinetic studies provide a simple model for the major steps
in the actin-activated hydrolysis of ATP by myosin. Do they provide
any insight into the mechanism of contraction? The criterion to be
applied in answering the question is that the enzyme scheme should
fit together with the postulated mechanical events in a natural way.
If we have to force the scheme to fit by arbitrarily introducing
extra steps or if it is necessary to assume that steps occurring in
the muscle have very different rates than in solution, then the
enzyme scheme is a poor model. As will be discussed below, it is
necessary to consider the constraints introduced by the lattice so
that a satisfactory mechanism cannot be derived from solution
measurements alone.

The main steps in the enzyme scheme are shown in Fig. 1 (b) for
comparison with the mechanical scheme. The four steps in the contractile cycle are (a) the dissociation of the actin-bridge complex,
(b) the movement of the free myosin bridge, (c) the recombination
of the bridge with actin, and (d) the drive stroke. In the enzyme
mechanism the steps are (a) the binding of ATP to AM and very rapid
dissociation of actin, (b) the splitting of ATP on the free myosin,
(c) the recombination of actin with M · ADP · P, and (d) the dissociation of the products from AM.

incubated 30 min with pyrophosphate (10^{-5} M) in 0.05 M KCl, 0.05 M
Tris buffer, pH 8, 2×10^{-3} M $MgCl_2$, 5°C. 0.8 ml mixed with 1.5 ml
actin (4×10^{-5} M) plus 10^{-3} M unlabeled pyrophosphate (-o-o-), or
pyrophosphate alone (-•-•-); column buffer contained 10^{-3} M pyrophosphate. Protein peak indicated by the arrow eluted 60 sec after
mixing.

TABLE 1

Rate Constants for Actin-HMM Mechanism

k_1	k_2	k_3	k_4	k_5	k_6	k_7
$2 \times 10^6 \ M^{-1} sec^{-1}$	$100-150 \ sec^{-1}$	$0.05 \ sec^{-1}$	$1 \times 10^6 \ M^{-1} sec^{-1}$	$\pm 1000 \ sec^{-1}$	$10^5 \ M^{-1} sec^{-1}$	$10-20 \ sec^{-1}$

Conditions: $0.05-0.1$ M KCl, $1-5 \times 10^{-3}$ M $MgCl_2$, pH 8, 20°C.

The similarities of the two schemes are obvious. Steps (a)
and (c) in both schemes involve the dissociation and recombination
of actin and myosin, and it appears quite reasonable to identify
the corresponding steps. The chemical mechanism provides the explanation of how a cycle involving dissociation and recombination
of myosin bridges with actin filaments could be coupled to ATP
hydrolysis. The myosin-product complex is a metastable state with
a long half life which can await the arrival of an actin unit which
is properly oriented to allow interaction.

How much farther can we push the similarities? It is natural
to adopt the view that either the binding of ATP or the splitting
of ATP [step (b)] induces a configuration change which accounts for
movement of the free bridge. One could go even further and suppose
that after formation of the AM · ADP · P complex the release of
products leaves the AM in a configuration rather like a stretched
spring, so that tension is exerted or movement is produced in
returning to the original configuration [step (d)].

This interpretation is not unreasonable and uses the popular
notion that the binding of a small molecule is capable of changing
the configuration of a protein. However, our kinetic studies provide no evidence for a configuration change of the myosin molecule
associated with ATP binding or hydrolysis. Furthermore, there is
no convincing evidence for a large scale shape change of myosin.
One might infer that changes in configuration occur but the kinetics
of such a process have not been measured and there is at present no
experimental justification for fitting configuration changes into
the kinetic scheme.

In the absence of positive evidence that interaction with substrate or products is the direct cause of bridge movement, the
interpretation of A. F. Huxley [8] is also compatible with the
kinetic scheme. Huxley proposed, as we do also, that ATP binding
causes dissociation, but the movement of the free bridge is due to
thermal vibrations. The available evidence does not allow a choice

to be made between the two mechanisms but there are difficulties
which prevent us from accepting a naive model of substrate induced
configuration changes.

If the rates measured in solution are a proper measure of
processes in the muscle there are objections to most enzyme models
which have been proposed. It has generally been assumed that activation by actin required hydrolysis to occur with the actin-myosin
complex intact, namely that AM $\cdot$ ATP $\longrightarrow$ AM + ADP + P is the step
in which energy is released and which is directly coupled to contraction. This leads to a very inefficient mechanism since dissociation of AM by ATP is a competing reaction, and in solution it
appears to be much faster than the steps involving hydrolysis or
product dissociation. Our scheme is free of this objection since
ATP binding causes dissociation and hydrolysis occurs on the free
bridge. However, there is a second objection which arises when we
compare actual values of rate constants with processes in the muscle.
At a high ATP and actin concentration the rate limiting step is
product dissociation from AM $\cdot$ ADP $\cdot$ P and the rate constant is the
order of $10-20 \ sec^{-1}$. This rate is sufficient to account for the
rate of energy liberation by a muscle. For example, the myosin
concentration is about $0.2 \ \mu M \ g^{-1}$, so the rate of hydrolysis would
be $2-4 \ \mu M \ g^{-1} sec^{-1}$. Unfortunately data are not available for rabbit
muscle; for frog sartorius at 0°C, the maximum rate is about 1 μM
$g^{-1} sec^{-1}$ [12]. The actual figure depends on what fraction of energy
turnover is due to the calcium pump. At 20°C the rate could be
$5-10 \ \mu M \ g^{-1} sec^{-1}$. Thus the chemical data appear to be able to
account for the rate of hydrolysis in muscle.

However there is a problem if the movement step (d) is equated
with a step in the chemical mechanism. The maximum velocity of
shortening of the rabbit muscles used to prepare the myosin is not
known but comparison with other fast muscles at 20°C, such as cat
or frog sartorius, suggests a rate of 5-10 lengths per second [1].
If the sarcomere length is 2 μ, the relative velocity of thick and
thin filaments is $5-10 \times 10^4$ Å per sec. The distance over which

the active bridge remains in contact with actin is hardly likely to
exceed 100 Å. Therefore the time to execute this movement is only
1-2 msec and we must associate with step (d) a maximum rate constant
of about 350-700 sec^{-1} which greatly exceeds the product dissociation rate.

We are forced to conclude from the rather scanty evidence that
movement or force generation is unlikely to be brought about by the
product dissociation step. But a simple model based on configuration
changes induced by substrate or products would have led us to expect
that dissociation of products drives the movement. In fact the
only step which is fast enough is the binding of ATP to AM. The
ATP concentration in the muscle is the order of 1-3 mM, consequently
the pseudo first-order rate for this step is 1000-2000 sec^{-1}.
Unless it is arbitrarily assumed that dissociation of bridges is
blocked in the muscle, then the kinetic evidence rules out this
possibility since at low speeds of shortening movement could not
compete against dissociation.

Furthermore, whatever mechanism is adopted for the enzyme
scheme the rate limiting step is still 10-20 sec^{-1} and is far too
slow to be equated with movement. Perhaps the lesson to be learned
is that a proper test of models requires a study of enzymatic and
mechanical properties on muscles from the same animal, presumably
the frog sartorius.

Nevertheless, the discussion does make clear the nature of the
problem and the direction for future studies. The discrepancy between the rates of movement and product dissociation is sufficiently
large that even taking into account the different sources of the
enzymatic and mechanical data the discrepancy is probably real. It
has been shown by Barany [1] that the ATPase activity of actomyosin,
which is a measure of the rate limiting step, correlates with maximum velocity of shortening. This important finding does not contra-
dict the conclusion regarding movement. The maximum velocity is
presumably determined by the rate of the contractile cycle which is
set by the rate limiting step of the enzyme mechanism.

Although it is not yet possible to give a detailed mechanism it can be asked what sort of scheme appears most reasonable and would provide a basis for the design of further experiments. The kinetic studies would appear to fit into a modified form of the A. F. Huxley scheme [8].

If we accept the discrepancy between the rate of movement and the rate limiting step in the enzyme mechanism the energy liberated in step (d) has to have some immediate source other than a step in the hydrolysis process. Huxley assumed that the free bridges are in thermal motion but are constrained to attach asymmetrically with respect to their equilibrium position. This could be attributed to the structure of the lattice although he did not specify the origin of the asymmetry. The bridge is treated as a spring which obtains elastic energy from Brownian motion while ATP hydrolysis supplies the energy necessary to drive the process in a cycle. The rate of movement is determined by the load both external and internal as long as this rate is small compared to the intrinsic rate of relaxation of the spring.

An alternative source of energy is the binding of actin to myosin. The attachment of myosin to actin sites in solution occurs at a 45° angle. In rigor, in insect flight muscle, the bridges are attached at this angle and point in the direction away from the Z line. This suggests that the energy of interaction is largest in the bent position and if the bridge attaches weakly in the straight position (an assumption) the source of energy is the bridge binding energy itself. The energy to dissociate the bent bridge is supplied by the binding of ATP.

The role of the bound products may then be to modulate the interaction and also to block the binding of ATP until the bridge reaches the bent position. For very rapid shortening products would not dissociate and the bridge would be broken by energy supplied by movement of other bridges. In this case, no ATP is bound or hydrolyzed and the energy dissipated in a cycle is approximately the energy necessary to bend a free bridge. In this model the globular

region of the bridge should be fairly rigid and energy must be provided to pivot the bridge at the base. Although this energy is lost
it has the function of returning the free bridge to the straight
position to ensure attachment at the beginning of the stroke.

Thus an attempt to supply a mechanism introduces assumptions
and speculations not directly related to the chemical evidence.
However, it indicates the kind of information that must be obtained
experimentally and emphasizes that it is the actual values of the
rate constants which determine the validity of a model.

The value of 10-20 sec^{-1} for the maximum rate of acto-HMM
hydrolysis of ATP has been questioned by A. Weber (in preparation),
who obtains 100 sec^{-1}. Since this value is the order of the hydro-
lytic step for free myosin, this step could be rate limiting and
the rate of product dissociation would be unknown and could be fast
enough to keep pace with bridge movement in the muscle. A. F. Huxley
and R. Simmonds [Nature, 233, 533 (1971)] have proposed a mechanical
model to account for rapid transient behavior in which the direct
source of energy is the binding of myosin to actin.

REFERENCES

[1] Barany, M., "The Contractile Process," A Symp. of New York
Heart Assoc., Little, Brown & Co., Boston, 1967.

[2] Eisenberg, E. and C. Moos, Biochem., 7, 1486 (1968).

[3] Eisenberg, E. and C. Moos, J. Biol. Chem., 245, 2451 (1970).

[4] Elliot, G. F., J. Theoret. Biol., 21, 71 (1968).

[5] Elliot, G. F., E. M. Rome, and M. Spencer, Nature (London),
226, 417 (1970).

[6] Finlayson, B., R. W. Lymn, and E. W. Taylor, Biochem., 8, 811
(1969).

[7] Finlayson, B. and E. W. Taylor, Biochem., 8, 802 (1969).

[8] Huxley, A. F., Prog. Biophys, 7, 255 (1957).

[9]   Huxley, H. E., J. Mol. Biol., 7, 281 (1963).

[10]   Huxley, H. E., Science, 164, 1356 (1969).

[11]   Huxley, H. E. and W. Brown, J. Mol. Biol., 30, 383 (1967).

[12]   Kushmerick, M. J. and R. E. Davies, Proc. Roy. Soc. (London) Ser. B, 174, 315 (1969).

[13]   Lymn, R. W. and E. W. Taylor, Biochem., 9, 2975 (1970).

[14]   Pringle, J. W. S., Prog. Biophys., 17, 1 (1967).

[15]   Reedy, M. K., K. C. Holmes, and R. T. Tregear, Nature (London), 207, 1276 (1965).

[16]   Sartorelli, L., H. J. Fromm, R. W. Benson, and P. D. Boyer, Biochem., 5, 2877 (1966).

[17]   Taylor, E. W., R. W. Lymn, and G. Moll, Biochem., 9, 2984 (1970).

[18]   Tonomura, Y., H. Nakamura, N. Kinoshita, J. R. Onishi, and M. Shigekawa, J. Biochem. (Tokyo), 66, 599 (1969).

[19]   Weber, A. and W. Hasselbach, Biochem. Biophys. Acta, 15, 232 (1954).

CHAPTER 4

PROBLEMS IN THE ANALYSIS OF FORCE-VELOCITY RELATIONS IN HEART MUSCLE*

Allan J. Brady

Department of Physiology and
Los Angeles County Cardiovascular Research Laboratory
University of California at Los Angeles
Center for Health Sciences
Los Angeles, California

*Study supported by Grant HE 09257-05 from the United States Public
Health Service.

I.   PROBLEMS IN THE ANALYSIS OF FORCE-VELOCITY
RELATIONS IN HEART MUSCLE

From a physiological standpoint the hyperbolic force-velocity
(P-V) relation* proposed by A. V. Hill in 1938 [9] has been the
dominant basic concept guiding studies of mechanical function of
muscle. At that time the important link between muscle energetics
and mechanical performance appeared to have been made such that the
non-linear relation between load and velocity was clearly a fundamental property of the active contractile process and not simply a
manifestation of a non-Newtonian type of viscosity. Although this
correspondence between mechanics and energetics has proven to be
much more complex than originally proposed, nevertheless, the
interrelationship of mechanical performance and energy output has
remained as the focal point of studies of mechano-chemical coupling
in muscle. Indeed, a major effort continues among muscle physi-
ologists to show that the hyperbolic P-V relation is a fundamental
property of all muscle.

Interest in the P-V relation of muscle, particularly heart
muscle, has persisted for at least three reasons. (a) If the
relation were capable of analytical definition then it should be
possible to identify the constants of the relation, a, b, P_o, V_{max},
and relate these to the fundamental energetics of heat production
and hopefully mechano-chemical processes whereby chemical energy is
transformed into mechanical energy. (b) The slow onset of the
cardiac active state and its extreme lability, i.e., its physiological inotropies, could give some clue as to the mechanism of E-C
coupling in terms of both tension and shortening capabilities.
(c) Clinically, it would be extremely useful if a method could be
devised whereby fiber length or cardiac volume related inotropies
could be differentiated from those in which a chemical modification

---

*$(P + a)(V + b) = (P_o + a)b = (V_{max} + b)a$ where P and V are muscle
force and shortening velocity, respectively. P_o and V_{max} are the
axial intercepts and a and b are experimentally determined constants.

in contractility has occurred. Because of the fundamental nature
of these questions it is surprising that the answers have been so
slow in resolution. The retarded progress, however, is better
appreciated when the complexities of the system are considered.

For example, the study by Abbott and Mommaerts [1] showed that
in afterloaded contractions of mammalian papillary muscle, shortening velocity was nonlinear with load over the range of loads that
could be studied beginning with a necessary preload. However, the
position of the curves depended upon the inotropic state of the
muscle and the preload, so that the important force-velocity factors
V_{max} and P_o, did not appear to be constant. Obviously, in order to
accurately evaluate the constants of Hill's equation and test their
behavior in the many cardiac inotropies, it is necessary to know
these coordinate intercepts. But several limitations in the complete description of the cardiac force-velocity relation have become
apparent, first from the work of Abbott and Mommaerts [1] and later
from others [2,7,12,15]. The major problems are as follows: (a)
Heart muscle must be preloaded in order to establish a rest length
at which active tension is measurable; the zero load velocity
(V_{max}), then, must be obtained by extrapolation (Fig. 1). As muscle
length is increased, by increasing the preload, this extrapolation
must be made over a greater range of forces and the value of V_{max}
becomes more inaccurate. Thus the length dependence of V_{max} is
difficult to establish simply from the velocity intercept. (b) On
the other axis, P_o is uncertain. In the Hill context P_o was independent of muscle length around in situ muscle length. In heart
muscle in the physiological range, maximum isometric twitch tension
is strongly length dependent (Fig. 1). In addition, isometric
tension can be grossly augmented in several ways. Thus, P_o for
heart muscle in the Hill concept is also ambiguous. (c) The dramatic
translation of the whole P-V relation with changes in initial muscle
length (Fig. 1) and alterations in the inotropic state of the muscle
leave the basic character of the P-V relation and especially its
constants, a and b, unclear.

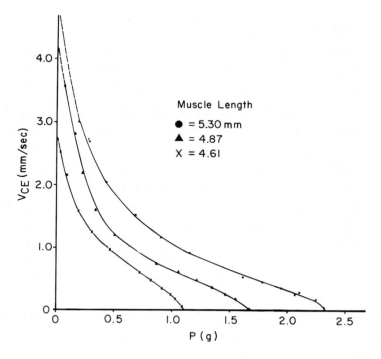

Figure 1. Examples of force-velocity data from afterloaded
contractions of rabbit papillary muscle. Dashed sections of curves
illustrate the arbitrary nature of the extrapolation to the velocity
axis especially with larger preloads.

More recently, studies of cardiac P-V relations reveal additional complexities which underlie these differences. First, in
afterloaded contractions the peak shortening velocity, particularly
with light loads, is not reached immediately, (Fig. 2) but takes
considerable time to develop. The delayed maximum cannot be attributed to inertial effects but suggests that maximal activation of
muscle contractility is delayed. Indeed, the quick stretch technique used to establish the early onset of active state in skeletal
muscle [10] when applied to cardiac muscle indicates that the
ability of heart muscle to bear a load is manifested little more
rapidly than normal twitch tension [3]. This opens the question as

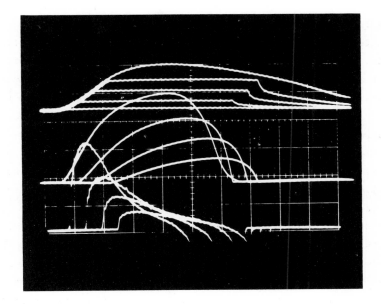

Figure 2. Examples of four afterloaded contractile responses
superimposed on an isometric contraction showing tension (upper
traces), shortening (middle traces), and shortening velocity (lower
traces). Notice (a) the abbreviation of the active state when
shortening occurs (tension and length traces) and (b) the relatively
slow rise of velocity to its peak value with each afterload.

to whether the maximum ability of the muscle to shorten is not also
delayed. In any case, in afterloaded contractions a substantial
and variable external shortening may occur before the peak velocity
is attained with each load. Thus, the length parameter, on which
contractile force is so strongly dependent, appears as an uncon-
trolled variable in these P-V measurements.

A second problem comes from the observation that cardiac muscle
is considerably more compliant than skeletal muscle (6% vs 2%).
Hence, before a load is lifted, considerable internal shortening
must occur which moves the functional capacity of the muscle further
to the left on the Starling length-tension relation. Obviously,

this internal shortening and consequent reduction in functional
capacity will increase, the greater the load [cf. Fig. 3 (b)].
Again, the nonconstant length factor would be expected to influence
P-V measurements. A third complication is that in the presence of
substantial resting tension, necessary to set initial muscle length
[Fig. 3 (a)], the parallel elastic element (PE) may introduce an
unknown contribution of force to the load. Thus, if the series
elastic element (SE) does not bear resting tension, then no change
in resting tension occurs during the isometric phase of the contraction but as afterloaded shortening begins PE force declines and
transfers its contribution to the contractile element (CE). If
peak shortening velocity does not always occur at the same muscle
length then the amount of resting tension transferred to the CE is
not constant for all loads. Hence, the true CE load may not be
known. On the other hand, if SE does bear resting tension, then
during the isometric phase of a contraction, PE tension declines as
CE shortens. The transfer of PE tension to the CE during this
period will vary depending on the afterload. Again, at the beginning of the afterloaded phase the true CE load is not known,
although it now remains constant throughout the isotonic shortening
period. Thus, resting tension is not a constant in each P-V analysis and must be accurately assessed, especially at longer muscle
lengths where the preload is high.

It is clear, then, that a mechanical analysis of heart muscle
requires the control or at least a knowledge of the four parameters,
force, velocity, length, and time. But still other interactions of
contractile parameters have been shown to occur. For example,
shortening appears to uncouple or terminate the active state of the
muscle. This is most apparent when the relatively short duration
of an isotonic contraction is compared with an isometric response
(Fig. 2). The muscle cannot bear a force equal to the afterload
nearly as long in an isotonic contraction as it can in an isometric
contraction. The uncoupling effect is most dramatic in the relaxation period; nevertheless, it does appear while the active state is
rising [4].

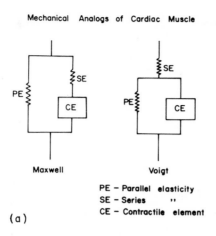

Mechanical Analogs of Cardiac Muscle

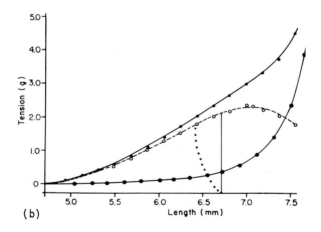

(a)

(b)

Figure 3.   (a) Mechanical analogs of papillary muscle.
(b) Length-tension characteristic of isometric contractions in
papillary muscle.   Lower solid curve, resting tension; upper solid
curve, total muscle tension; dashed curve, active tension; vertical
solid line at 6.7 mm, tension development of CE at constant CE
length; dotted line, estimated length-tension course of CE during
an isometric construction with muscle fixed at 6.7 mm length ($\Delta \ell_{SE}$
assumed to be 5% of muscle length).

From these considerations it is clear that force-velocity
relations derived from afterloaded contractions in heart muscle are
essentially meaningless with regard to the establishment of fundamental mechano-chemical relationships. Aside from the geometrical
complexity of the tissue, the contractile parameters are simply
much too interactive to allow formulation of any basic tenets from
a simple measurement of force and velocity alone. A more rigorous
analysis of all muscle parameters is demanded.

II. METHODS OF FORCE-VELOCITY MEASUREMENT IN PAPILLARY MUSCLE

Several methods have been used to calculate P-V relations
whereby CE length changes are taken into account during the isometric
and isotonic phases of contraction. The method used by Edman and
Nilsson [7] involves a quick release of the muscle from an isometric
state to various loads at a given time during selected contractions.
A rapid undamped shortening occurs first as SE tension falls to
match the new load. Then the slower CE shortening continues with
the attached load. Releases are made to a series of different loads.
When a family of length-time and velocity-time curves for these
responses are superimposed the velocity at a time after release can
be determined for each load. The velocity is then plotted as a
function of load. In their experiments CE length at which the ve-
locities were measured varied by less than 1% ℓ_o. In this analysis
muscle length is taken as a direct measure of CE length since the
release is given at the same time in each contraction and hence at
the same SE extension. The rapid change in length accompanying the
release to the load is assumed to occur only in SE since CE cannot
shorten more rapidly than V_{max}. Any subsequent change in muscle
length must then occur in CE alone unless PE bears significant
tension.

Obviously uncoupling effects accompanying the quick release
and the restriction to low resting tensions limit the usefulness of
the technique. Also, even a 1% variation in CE length can significantly modify the contractile response. However, Edman and Nilsson

[7] obtained reasonably good hyperbolic P-V relation for rabbit papillary muscles under these conditions. But, the calculated V_{max} was strongly dependent upon the initial muscle length as well as time after activation.

Pollack [13] arrived at a similar conclusion after correcting CE length for SE characteristics using the early P-V data of Sonnenblick [15] and SE data obtained from later experiments. Noble et al. [12] have made similar measurements, correcting for CE length changes and load changes in terms of various models. In general, when CE length is accurately considered in terms of CE load, the resulting P-V relations derived from afterloaded contractions are nonlinear and approximately hyperbolic. On the other hand, both V_{max} and P_o appear to be strong functions of both time and CE length. Pollack, Edman, and Noble all found that the highest V_{max} occurred late in the rising phase of the contraction.

The more recent studies of Brutsaert and Sonnenblick [6] report a hyperbolic P-V relation in cat papillary muscle where SE length corrections are applied only to P_o. With only this correction V_{max} becomes independent of initial muscle length in the vicinity of the peak of the Starling relation. The justification for the omission of SE length corrections on lighter loads lifted early in the contraction (near V_{max}) is their assumption that the characteristic shortening ability for each load has an early rapid onset and maintains a plateau of activity until the "active state" begins to decline at a time somewhat before P_o is reached. The basis for this assumption comes from their observation that shortening velocity is set by the load and CE length alone. However, underlying these studies of afterloaded shortening is the fact that as the active state is rising, CE is shortening against either the SE or the attached load. The resultant shortening velocity of CE will have its maximum value wherever the rate of rise of the active state equals the rate of loss of myofilament interaction as CE moves down the Starling curve. This value can occur at most any time and at most any CE length depending on the load and the starting muscle

length. Thus the peak shortening velocity with which a load is
lifted is not necessarily the "characteristic" velocity of the CE
for the load in question. Conclusions regarding P-V relations and
cardiac energetics based on data obtained from afterloaded contractions, therefore, may be grossly misleading. Only a complete analysis of CE length, force, and velocity, as functions of time, can
reveal the basic nature of the P-V relationship.

III. CONTROLLED CONTRACTILE ELEMENT LENGTH ANALYSIS

Since both the afterloaded shortening and quick release techniques introduce either time and CE length or uncoupling complexities
into the P-V analysis it is desirable to find a technique in which
the CE force, velocity, length, and time parameters can be measured
as directly as possible and preferably where uncoupling effects are
minimized, i.e., CE velocity is zero or, at least, controlled. The
following is an extension of the controlled stretch "active state"
method [5] which minimizes most of these problems.

Before any calculation of CE force as a function of muscle
length can be made, it must be determined whether SE bears resting
tension and thus whether the initial SE length is influenced by
initial muscle length. This is done by applying small length perturbations (< 1%) to the passive and actively contracting muscle at
various times during successive contractions. The change in tension
(ΔP) due to the small stretch or release is a measure of the change
in muscle stiffness as active tension develops. If the perturbations
are sufficiently rapid (faster than CE can change length) and low in
amplitude then the change in muscle stiffness from rest can be
attributed to the SE for either analog. When stiffness ($\Delta P/\Delta l$) is
plotted against the tension at which the perturbation was applied
the resulting curve characterizes both resting stiffness (the
initial point) and SE (Fig. 4). When the initial muscle length is
increased, however, the resulting stiffness curves are translated
differently depending on the analog. If SE does not bear resting

tension, then these curves are translated upward and to the right
in accordance with the PE relation alone. Since SE always starts
from its zero tension length in this model, the slope of the curve
will be the same at longer muscle lengths. Thus the curves will be
separated in accordance with the change in PE stiffness.

If SE does bear resting tension, then as initial muscle length
is increased, SE is also stretched so that resting stiffness represents that of the series combination of SE and PE; but as soon as
CE is active the increase in muscle stiffness always lies on the
original SE curve, but starting at a higher level of tension. Thus,
with an increase in muscle length the stiffness curves overlap. Of
course, if neither analog is appropriate then some different translation will occur at longer lengths and the muscle cannot be used
for further analysis unless the more complex analog can be uniquely
defined.

Once the SE stiffness curve is known, it can be integrated, at
least graphically, so that the SE length-tension relationship is
known. From this and a measurement of muscle length, changes in CE
length can be calculated. It can be seen that this method of SE

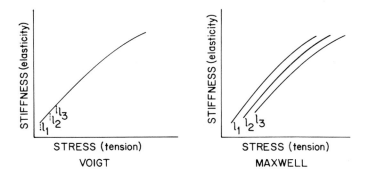

Figure 4. Sketch of stiffness-stress responses of Voigt and
Maxwell analogs at different initial muscle lengths, ℓ_1, ℓ_2, ℓ_3.
Curves are separate and parallel in Maxwell model but superimposed
in Voigt model.

analysis is not as limited by ambiguities of possible CE uncoupling as are quick releases to various loads; nor does the method assume that the SE tension-extension relation is necessarily exponential.

The PE length-tension relation can be obtained directly by simply applying a ramp stretch to the muscle if PE alone bears resting tension; if PE and SE are in series then PE is calculated from the known SE relation and the series combination of SE and PE in the resting muscle, i.e., $1/PE = (1/P_t) - (1/SE)$ where P_t represents the length-tension relation for the series combination of PE and SE (resting length-tension relation). We have neglected the known creep phenomenon of the resting muscle but this complication can be minimized by measuring the resting length-tension relation with ramps of the approximate velocity with which shortening or stretch occurs in the active muscle.

Now for those muscles which can be represented by one or the other of the analogs, a total description of their length, tension, and velocity characteristics can be deduced as functions of time. This method involves applying a controlled stretch to the muscle during contraction [5] such that CE length is maintained constant or constrained to move at a known velocity throughout the contraction. This is accomplished most simply for the model in which SE bears resting tension. The SE length-tension relation is first determined by integration of the stiffness data obtained from the small stretch and release technique as indicated above. Then this SE relation forms a control function whereby the muscle is stretched by an amount determined by this function, as active tension is developed by the muscle (Fig. 5). In other words, as the muscle develops tension during a twitch, SE is stretched by a muscle puller continuously and just sufficiently to prevent CE from shortening. This is the V = 0 state. Now if a ramp signal is added (electronically) to the control function signal then CE follows this ramp signal, i.e., CE velocity is a constant equal to the slope of the ramp since the control function, in effect, cancels out SE. Obviously, the ramp may be positive or negative going. Now, since

CE velocity is set, CE length is known continuously from the ramp, and total muscle force is known as a function of time. Thus the force-velocity-length relation of CE can be determined by releasing the muscle with various ramps and by starting at different initial muscle lengths.

One correction must be kept in mind. Total tension is CE force plus PE tension. If the experiment is performed at short muscle lengths where PE tension is low and shortening ramps are used, then CE tension is approximately equal to total muscle tension. At longer muscle lengths or when a stretch ramp is used then the contribution of PE must be evaluated. This can be done as follows. The control function curve describes the length-tension relation of SE. The resting length-tension relation describes the series combination of PE and SE. The difference in the reciprocal of total tension relation and the SE relation is the reciprocal of the PE

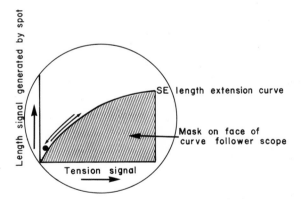

Figure 5. Sketch showing generation of control function for active state measurement and P-V measurements in controlled stretch experiments. Circle depicts face of oscilloscope used as a curve follower. Mask is made from SE length-extension relation. Tension signal from muscle drives spot along x-axis of oscilloscope but spot is constrained to follow edge of mask by a photomultiplier on which spot is focused. (The photomultiplier drives the y-axis amplifier of the scope.) The desired length signal, i.e., the control function, is then derived from y amplifier of scope.

relation. Since overall muscle length and SE length are known during the ramp, PE length is known; hence PE tension can be calculated
and thus CE force.

IV. FORCE-VELOCITY RELATIONS FROM ISOMETRIC CONTRACTIONS

Since the CE transfers its force to the ends of the muscle
through SE, CE must shorten somewhat even during isometric contraction (Fig. 3). It faces a complex load, however, because SE behaves
like a nonlinear spring, i.e., it becomes stiffer with stretch. But
if the SE length-tension characteristic is known, then CE length as
well as velocity can be calculated [8]. This is done as follows:

$$\ell_{CE} = \ell_m - \ell_{SE} \tag{1}$$

hence

$$\frac{d\ell_{CE}}{dt} = V_{CE} = \frac{d\ell_{SE}}{dt} \tag{2}$$

where ℓ_{CE}, ℓ_m and ℓ_{SE} are the extensions of the CE, the muscle and
SE respectively from their reference rest lengths. V_{CE} is CE
velocity and V_m is muscle velocity (equal to zero for an isometric
contraction).

Now

$$\frac{d\ell_{SE}}{dt} = \left(\frac{dP_{SE}}{dt}\right)\left(\frac{dP_{SE}}{d\ell_{SE}}\right) \tag{3}$$

since the SE length-tension relation is continuous over the length
interval in question.

Therefore, if dP_{SE}/dt, i.e., ($\dot{P}_{SE}$) is measured and SE stiffness
$dP_{SE}/d\ell_{SE}$ is known as a function of P_{SE}, then V_{CE} can be calculated
throughout the twitch. Again it must be established which analog
describes the muscle so that $\dot{P}_{SE}$ can be determined, i.e., in the
Maxwell case $\dot{P}_{SE} = \dot{P}_t - \dot{P}_{PE}$ and in the Voigt case $\dot{P}_{SE} = \dot{P}_t$.

To obtain V_{CE} as a function of ℓ_{CE} in the Voigt case, for example, ℓ_{CE} is plotted against P during the twitch ($\ell_{CE} = 1 - \ell_{SE}$) at several different initial muscle lengths. V_{CE} is also plotted against P for the same muscle lengths. Then, at given P's, V_{CE} and ℓ_{CE} can be picked off from these plots and V_{CE} plotted against ℓ_{CE} as a family of curves for different P. Thus the trend in velocity-length relations can be established as tension rises in the muscle as a function of time. Furthermore, at given ℓ_{CE}'s, V_{CE} as a function of P can be read off the family of curves so that P-V relations at different CE lengths can be determined.

Both the controlled stretch method [16] and the calculation of CE velocity during isometric or isovolumic contractions [11,14] are currently in use. In the later studies SE is assumed to be exponential, including the left heart analysis, but this assumption is not always valid for the full range of papillary muscle forces and has not been established for the organ.

V. SUMMARY

The study of force-velocity relations in heart muscle is considerably more complex than in skeletal muscle because cardiac P-V measurements cannot be made in a domain of muscle function where length and time are not inherent variables. Additionally, resting tension contributions must be apprised and, with each perturbation of muscle function we must evaluate possible myofilament uncoupling effects. While each technique may obviate or minimize one parametric complication it may introduce another. Assuming mechanical homogeneity of the muscle preparation, the controlled stretch technique (with positive and negative ramps superimposed) appears to offer the most relevant analysis although the technique is complicated and demands precision control of all contractile parameters. The analysis of force-velocity relations in the isometric state is also promising provided all parameters are considered. However, the isometric contraction represents a rather limited expression of contractile function which restricts its utility as means of fully characterizing cardiac muscle contractility.

In perspective, it can be seen that as more of the cardiac contractile parameters are systematically considered the promise of new insight into the mechano-chemistry of cardiac muscle becomes realistic. We have yet to face the complexities of analysis introduced by myocardial mechanical inhomogeneities but now with methods to control the known parameters of contraction we are in a position to evaluate these additional structural complexities in an orderly fashion.

REFERENCES

[1]   Abbott, B. C. and W. F. H. M. Mommaerts, "A study of inotropic mechanisms in the papillary muscle preparation," J. Gen. Physiol., 42, 533-551 (1959).

[2]   Brady, A. J., "Time and displacement dependence of cardiac contractility: Problems in defining the active state and force-velocity relations," Federation Proc., 24, 1410-1420 (1965).

[3]   Brady, A. J., "Onset of contractility in cardiac muscle," J. Physiol., 184, 560-580 (1966).

[4]   Brady, A. J., "Mechanics of isolated papillary muscle," in Factors Influencing Myocardial Contractility (R. D. Tanz, F. Kavaler, and J. Roberts, eds.), Academic, New York, 1967, pp. 53-64.

[5]   Brady, A. J., "Active state in cardiac muscle," Physiol. Rev., 48, 560-600 (1968).

[6]   Brutsaert, D. L. and E. H. Sonnenblick, "Force-velocity-length-time relations of the contractile elements in heart muscle of the cat," Circ. Res., 24, 137-149 (1969).

[7]   Edman, K. A. P. and E. Nilsson, "The mechanical parameters of myocardial contraction studied at a constant length of the contractile element," Acta Physiol. Scand., 72, 205-219 (1968).

[8]   Hefner, L. L. and T. E. Bowen, Jr., "Elastic components of cat papillary muscle," Am. J. Physiol., 212, 1221-1227 (1967).

[9] Hill, A. V., "The heat of shortening and the dynamic constants of muscle," Proc. Roy. Soc., B126, 136-195 (1938).

[10] Hill, A. V., "The abrupt transition from rest to activity in muscle," Proc. Roy. Soc., B136, 399-520 (1949).

[11] Hugenholtz, P. G., R. C. Ellison, C. W. Urschel, I. Mirsky, and E. H. Sonnenblick, "Myocardial force-velocity relationships in clinical heart disease," Circ. Res., 51, 191-202 (1970).

[12] Noble, M. I. M., T. E. Bowen, Jr., and L. L. Hefner, "Force-velocity relationship of cat cardiac muscle, studied by isotonic and quick-release techniques," Circ. Res., 24, 821-833 (1969).

[13] Pollack, G. H., "Maximum velocity as an index of contractility in cardiac muscle," Circ. Res., 26, 111-127 (1970).

[14] Ross, J., Jr., J. W. Covell, E. H. Sonnenblick, and E. Braunwald, "Contractile state of the heart characterized by force-velocity relations in variably afterloaded and isovolumic beats," Circ. Res., 18, 149-163 (1966).

[15] Sonnenblick, E. H., "Force-velocity relations in mammalian heart muscle," Am. J. Physiol., 202, 931-939 (1962).

[16] Brady, A. J. (unpublished).

CHAPTER 5

MECHANISMS OF GROWTH AND ATROPHY OF SKELETAL MUSCLE

Alfred L. Goldberg

Department of Physiology
Harvard Medical School
Boston, Massachusetts

I. INTRODUCTION

During the past several years, my laboratory has been studying
the cellular mechanisms underlying growth and atrophy of skeletal
muscle. Our initial goal was to elucidate the biochemical events
through which use and disuse influence muscle size. Although much
progress has been made toward characterizing work-induced hypertrophy or disuse atrophy, we are still a long way from understanding
these processes. In the course of these studies, however, a number
of new and unexpected findings have been made concerning the control
of growth, amino acid, and protein metabolism in this tissue. In
addition to reviewing our findings on work-induced growth, the

present manuscript attempts to summarize some of these recent
developments, which in our opinion affect certain widely accepted
views of skeletal muscle as a tissue. The following areas will be
discussed: (a) biochemical events occurring during compensatory
growth and the relationship of hypertrophy to other growth processes,
(b) effects of exercise on amino acid transport into muscle, (c) new
findings on amino acid degradation in muscle, and (d) the importance
of protein breakdown for the control of muscle size.

II. WORK-INDUCED GROWTH OF SKELETAL MUSCLE

Like most organs of the body, muscle changes in mass with
changes in physiological demand; increased work leads to rapid
hypertrophy, while decreased work causes atrophy. These adaptive
responses [27] are probably of important selective advantage to the
organism and enable it to acquire new skills and to compensate for
disease or injury (e.g., cardiac hypertrophy). Before approaching
the biochemical events responsible for compensatory hypertrophy, it
was necessary to define more precisely what cellular changes occur
during this process and to clarify the relationship between work-
induced hypertrophy and other growth processes in muscle.

An essential first step for such investigations was the development of a simple and highly reproducible experimental system
for inducing hypertrophy of skeletal muscles [12]. In the hind
limb of the rat, three muscles act synergistically to extend the
ankle and thus support body weight: the gastrocnemius, the soleus,
and the plantaris. When the gastrocnemius portion of the Achilles
tendon is sectioned on one limb, the remaining plantaris and soleus
muscles must alone support the body weight on that side and undergo
rapid compensatory growth. The contralateral limb received a sham
operation, and the corresponding muscles of this limb served as
controls. Following tenotomy of the gastrocnemius, the soleus and
plantaris muscles on that side grew at a markedly accelerated pace,
while the muscles of the control limb grew at their normal rate

(Fig. 1). Within five days after operation, the soleus of the
tenotomized limb was 40% heavier than its contralateral control and
the plantaris 20% larger. The great advantage of this experimental
technique is that it affords the opportunity of comparing a hypertrophying and a nongrowing control organ within the same animal.

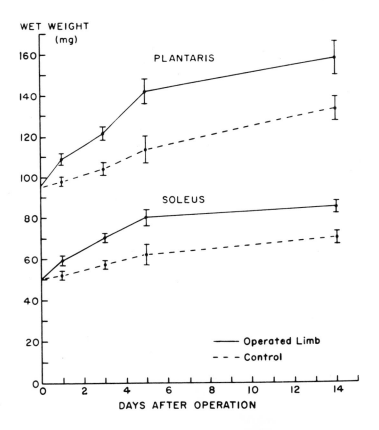

Figure 1. Sizes of soleus and plantaris muscles after section
of the gastrocnemius tendon in normal rats. During the first five
days, the muscles of the tenotomized limb show an accelerated growth
/rate presumably in response to their increased work load/ Each
point represents the average of five animals. Standard errors are
indicated.

Thus animal-to-animal variations (e.g., in diet, hormonal content, etc.) are minimized, and as a result a number of experimental questions can be investigated more precisely than was possible heretofore.

A most surprising finding was the rapidity of the work-induced growth process. /Twenty-four hours after operation, increase in mass was already apparent, and by the end of five days, it was essentially complete. After this time, the muscles of the two limbs continued growing at about the same rate, in line with overall body growth. Similarly rapid hypertrophy was also found when analogous experiments were performed on adult animals in which total body growth was much slower./ Such rapid changes in muscle size had not been anticipated by us simply because the capacity of muscle to undergo compensatory growth (in contrast with that of liver and kidney) has not been generally recognized. It is of interest that recent studies of cardiac hypertrophy, induced by artificial coarctation of the aorta or by other stimuli, have also found that cardiac muscle can increase in weight by 40-50% within a week [1,10].

The gain in weight during compensatory growth (Fig. 1) reflects an increase in muscle protein (Table 1). The hypertrophying muscle incorporated radioactive amino acids into protein more rapidly than controls, and the total increase in incorporation of ^{3}H-leucine during hypertrophy was found to be directly proportional to the change in muscle weight [14]. The conclusion of increased protein synthesis received further support from the observations by others that cell free extracts from hypertrophying skeletal or cardiac muscles were more active than controls in incorporating amino acids into protein [30,42,52].

The finding of greater protein synthesis also correlated with an increased RNA content in the hypertrophying soleus [23]. The RNA concentration was maximal six days after tenotomy of the gastrocnemius and declined to control levels following the period of rapid growth. Treatment with actinomycin D, an inhibitor of RNA synthesis, blocked the increase in muscle RNA and blocked work-

TABLE 1

Biochemical Changes During Hypertrophy of Soleus Muscle[a]

Conclusions	Comment
1. Increased amino acid transport	^{14}C-AIB uptake increased within 3 hr.[b]
2. Increased protein synthesis	^{3}H-Leucine incorporation increased within 6 hr.[b] Maximal on day 3.[b]
3. Increased RNA synthesis	Incorporation of ^{3}H-orotic acid increased within 6 hr.[b] and maximal on day 3.[b]
4. Increased protein content	Evident within 36 hr.[b] Maximal on day 5.[b] Soluble, myofibrillar, and collagenous protein all increase.
5. Decreased protein breakdown	Especially sarcoplasmic proteins.
6. Increased DNA	DNA content and number of nuclei increased but fiber number constant.
7. Increased ^{3}H-thymidine incorporation	Primarily in nonmuscular cells.

[a]All comparisons were made between contralateral hypertrophying and nongrowing controls.

[b]Time after tenotomy of synergistic gastrocnemius muscle.

induced growth [18]. In addition, the hypertrophying soleus was found to incorporate ^{14}C-orotic acid, a precursor of uracil, into RNA more actively than the contralateral control. Thus the increase in RNA appears to result from greater RNA synthesis and appears to be an essential part of the growth process.

Both sarcoplasmic (soluble) and myofibrillar (contractile) proteins increase in absolute amount during hypertrophy, although the proportion of soluble proteins increase disproportionately [20].

This latter finding was unexpected and as yet we have no idea of the physiological significance of this change in the relative proportions of these two classes of proteins. Histological studies of the contralateral muscles demonstrated that this compensatory growth occurred through an increase in the mean diameter of the muscle fibers without any change in their number. Nevertheless, the hypertrophied muscle was found to contain a greater number of nuclei [12] and a greater amount of DNA than the contralateral control [30]. Direct measurements of the incorporation of ^{3}H-thymidine further indicated new DNA synthesis within the hypertrophied muscle. Although these findings at first suggest the somewhat heretical possibility of division of muscle cells, recent autoradiographic studies have indicated that the majority of the labeled nuclei were of connective tissue origin. Grove et al. [29] and Morkin and Ashford [41] have also found increased DNA synthesis by nonmuscular cells during hypertrophy of the heart. Thus, in both types of muscle, increased work leads to hypertrophy of muscle cells and concomitant hyperplasia of the stromal components.

The physiological significance of this proliferation of connective tissue cells is unclear, since we have no idea of the function of the nonmuscular components. Interestingly the collagen content of the soleus also increased during hypertrophy, out of proportion to the increase in remaining muscle proteins. (Very similar observations have been made by Bartsova et al. [3] and by Buccino et al. [6] in hypertrophied hearts.) This increased collagen content could have profound effects on the mechanical characteristics of the muscle. In addition, it is possible that the increased synthesis of collagen represents a compensatory response of the connective tissue cells, permitting the hypertrophied muscle to exert increased force.

III. ENDOCRINE REQUIREMENTS FOR MUSCULAR HYPERTROPHY

A number of studies were undertaken to clarify the relationship between work-induced growth of muscle and the increase in muscle

size that occurs during normal development. It has long been known
that developmental growth requires the presence of pituitary growth
hormone [33,53]. After hypophysectomy, muscle growth stops abruptly,
whereas treatment with growth hormone reinitiates the growth process
[5,9]. To determine whether work-induced growth of muscle also
required the presence of pituitary hormones [12], we performed the
same operation in hypophysectomized rats. In these animals total
body growth was arrested following removal of the pituitary. Nevertheless, the soleus and plantaris of the operated limb underwent
compensatory growth in a manner similar to that seen with normal
animals. Following the five-day period of rapid hypertrophy, the
soleus and plantaris muscles of the hypophysectomized animals did
not grow further. Thus in these rats, compensatory growth has been
distinguished from normal developmental growth. When hypertrophy
is expressed as the ratio of the weights of the rapidly growing
muscles to that of their contralateral controls, the extent of such
growth was indistinguishable in normal and hypophysectomized animals
[12].

These experiments thus demonstrated that pituitary growth
hormone is not essential for work-induced hypertrophy. Like growth
hormone, insulin is believed to be essential for normal growth of
skeletal muscle. In its absence, protein synthesis in muscle is
significantly reduced [38], and muscle wasting becomes evident. On
the other hand, insulin added in vitro to the heart or to skeletal
muscle of diabetic animals causes a stimulation of protein synthesis
[34,54]. In addition, insulin appears essential for the anabolic
effects of pituitary growth hormone [32] and even for the differentiation of muscle cells from embryonic precursors [8].

Similar experiments to those described above were carried out
to determine whether insulin is also required for work-induced
hypertrophy [15]. Normal rats were treated with alloxan to make
them severely diabetic. In these animals, total body growth ceased,
and normal increase in muscle mass did not occur unless they were
administered large doses of insulin. Nevertheless, tenotomy of the

gastrocnemius in such rats caused rapid hypertrophy of the soleus
and plantaris. In fact, the average percent hypertrophy in the
diabetic animals was similar to that observed in controls. These
experiments thus indicate that work-induced hypertrophy is also
independent of insulin. Other studies indicate that increased
muscular activity has effects similar to insulin in promoting protein synthesis [14] and amino acid transport (see below).

In recent experiments [23] we have even shown that the compensatory growth process can occur in animals deprived of food and
thus in negative nitrogen balance. Tenotomy of the gastrocnemius
was performed in a series of rats two days after food was removed
from their cages. During the subsequent five days, the animals
decreased in weight by up to 35%, including a proportionate decrease
in total muscle mass. Nevertheless, the soleus and plantaris of
the tenotomized limbs showed an absolute as well as a relative
increase in weight. Thus the organism, while depleting muscle
protein reserves, appeared to permit the overworked muscles to
increase in size and to compensate for the increased demand upon
them.

Together these findings suggest that a definite hierarchy
exists among the various factors regulating muscle growth. Compensatory hypertrophy clearly takes precedence over hormonal growth
and even over endocrine signals for muscle depletion. The significance to the organism of work-induced growth and of developmental
growth are clearly different. Compensatory hypertrophy is specific
to the organ involved and occurs more rapidly than the growth
process that normally accompanies development. The latter process
takes place in many organs simultaneously and therefore probably
must depend upon hormones to integrate the anabolic responses of
the various tissues. Under many conditions, it would appear of
selective advantage to the organism to give greater priority to the
compensatory growth of an overworked muscle or heart, or to other
types of adaptive growth (e.g., wound healing) than to normal body
growth.

The differences demonstrated between work-induced growth and developmental growth in muscle are summarized in Table 2.

TABLE 2

Requirements for Growth of Muscle

Developmental growth	Work-induced hypertrophy
Pituitary hormones (growth hormone, TSH, etc.)	Occurs in hypophysectomized (nongrowing) animals
Insulin	Occurs in diabetic (nongrowing) animals
Adequate diet	Occurs in food-deprived animals (despite general muscle wasting)

Other studies have indicated that pituitary growth hormone by itself can induce growth in the absence of muscular work (i.e., in a denervated muscle) [18]. An important problem for future study is how these two types of growth differ in their biochemical and physiological consequences. Both involve increased protein synthesis in muscle, although they affect the composition of the tissue in distinct fashions [16]. Recent unpublished studies also indicate that both the hormone and increased work promote DNA synthesis in muscle.

These findings, obtained in carefully controlled and therefore somewhat artificial conditions, probably have overemphasized the separateness of these two growth processes. There is appreciable clinical experience suggesting that "normal growth" requires some muscular exercise. Thus one component of the developmental process may be work-induced and associated with increased activity of the growing organism. A major problem for future study is to determine how hormonal factors and exercise interact in the determination of size. Some studies in this direction have been initiated [23] and suggest that different endocrine factors (e.g., growth hormone and glucocorticoids) interact with muscular work in distinct fashions.

IV.   EFFECTS OF EXERCISE ON AMINO ACID TRANSPORT INTO MUSCLE

In skeletal muscle, as in certain other tissues, the rate of
protein synthesis appears to correlate with the rate of entry of
amino acids into the tissues [7,48]. For example, muscles from
hypophysectomized animals show a reduced ability to concentrate
amino acids. Similarly, insulin, which promotes protein synthesis,
also promotes amino acid transport into cardiac and skeletal muscle
through a direct effect on the muscle membrane [38] while cortisone,
which inhibits protein accumulation in muscle, also decreases the
entrance of amino acids [35]. We have carried out a number of
studies which indicate that in addition to these endocrine factors,
amino acid transport into muscle is influenced directly by the level
of muscular work.

These findings developed from studies of the transport of amino
acids into muscle during hypertrophy and atrophy [18,19]. In order
to study transport independently of amino acid incorporation, our
experiments employed the nonmetabolized amino acid analog, α-amino
isobutyric acid (AIB), which is accumulated by muscle by the same
transport system as the neutral amino acids. When ^{14}C-AIB was
injected intravenous, the hypertrophying soleus and plantaris muscle
accumulated AIB more rapidly and achieved a greater maximal distribution ratio (i.e., the intracellular/extracellular concentration
gradient) than did the contralateral nongrowing muscles. The
greater AIB accumulation could not be accounted for simply by an
increased blood flow to the hypertrophying muscle since differences
in AIB uptake were also seen when hypertrophying and control muscles
were removed from the body and incubated in vitro. Following the
period of rapid growth, amino acid transport by the soleus returned
slowly to control levels.

To examine the effects of decreased muscular work, related
studies were carried out on muscles atrophying because of denervation or disuse [19]. Section or anesthesia of the sciatic nerve or
section of the spinal cord reduced transport of amino acids into

the soleus and plantaris within 3 hr. Such changes were specific
to the paralyzed muscles. These experiments and the data on hypertrophying muscles led us to suggest that the level of muscular
activity is a major determinant of amino acid transport.

We have recently obtained more direct evidence for a relationship between muscular activity and amino acid transport through
experiments on isolated skeletal muscles stimulated to work in
vitro. The various changes in amino acid transport described above
all occurred as rapidly as could be studied effectively with the in
vivo approach. Experiments in collaboration with Dr. Charles
Jablecki and Mrs. Susan Martel were undertaken on isolated muscle
to investigate the earliest effects of increased muscular work in a
more controlled fashion than could be achieved with the intact
animal [25]. These studies employed the rat diaphragm, maintained in
Krebs-Ringer bicarbonate buffer and stimulated to contract electrically. In these experiments, both hemidiaphragms were removed
from the rat and suspended in an experimental chamber. One side
was stimulated directly for varying periods of time, while the
contralateral hemidiaphragm served as a control. Following this
period of exercise, both muscles were incubated with ^{14}C-AIB, and
the rate of transport of AIB compared.

When the diaphragm was stimulated at a rate of two or more
shocks per second (approximately the rate of respiration of the rat),
it subsequently accumulated AIB more rapidly than the unexercised
control (Fig. 2). This difference was evident 1/2 hr after stimulation and lasted for several hours. As was found in the in vivo
studies, the transport increased even when protein and RNA synthesis
in the muscle were prevented. The actual increase of AIB transport
was found to be a function of the amount of muscular work. As
shown in Table 3, stimulation at greater frequencies for a specific
time increased the magnitude of this effect. In addition, longer
periods of stimulation at a given frequency led to greater uptake.
However, any simple relationship between the extent of AIB accumulation and the number of contractions could not be demonstrated.

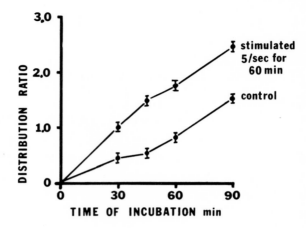

Figure 2. Effects of muscle stimulation on accumulation of
[14]C-AIB by isolated rat diaphragm from hypophysectomized rats. One
hemidiaphragm was stimulated repetitively at the rate of 5/sec for
1 hr. AIB uptake was then compared in the stimulated muscle and
the contralateral control. The differences in rate of accumulation
are highly significant ($p < 0.001$). Each point is the mean ± S.E.M.
of at least five observations. Distribution ratio was taken as the
intracellular/extracellular concentration ratio. Extracellular
space was defined by the ^{3}H-inulin space.

In general, if a given number of contractions occurred in rapid
succession, they appeared to have greater effect than if they occurred more slowly.

These experiments thus demonstrate a role of muscular activity
in regulating amino acid transport. Arvill [2] has also demonstrated increased AIB uptake upon electrical stimulation of the levator
ani muscle of the rat, while Lesch et al. [36] have observed increased transport in cardiac muscle fibers subjected to passive
stretch. Theoretically, this effect could result from the increased
contractile work or simply from the increased frequency of muscle
action potentials or from the increased release of acetylcholine in
the skeletal muscle. To distinguish between these possibilities,

TABLE 3

Effects of Electrical Stimulation on AIB Uptake by Rat Diaphragm[a]

A. Stimulation (at 5/sec) for varying times		
Duration of stimulation (min)	Difference between distribution ratios (stimulated muscle control)	P
3	0.01 ± 0.10	
12	0.33 ± 0.10	< 0.05
24	0.39 ± 0.10	< 0.02
40	0.68 ± 0.14	< 0.01
60	0.82 ± 0.08	< 0.001
B. Effect of varying stimulation frequency		
Stimulation frequency (pulse/sec)	Difference between distribution ratios (stimulated muscle control)	P
1.0	0.14 ± 0.05	< 0.05
2.5	0.87 ± 0.11	< 0.001
5.0	0.82 ± 0.08	< 0.001
7.5	0.78 ± 0.06	< 0.001
10.0	1.09 ± 0.06	< 0.001

[a]These studies were performed in the same manner as in Figure 2. Following the period of stimulation, the muscles were incubated for thirty minutes, and intracellular AIB content measured. Each point represents the mean ± S.E.M. of six rats.

both halves of the diaphragm were stimulated electrically at the same rate. One hemidiaphragm was left free to contract unopposed, while the contralateral muscle was stretched so that it contracted isometrically. In all seven experiments, the muscle contracting isometrically was found to concentrate AIB more rapidly [25]. An important problem for future work is the cellular mechanisms through which contractile activity influences amino acid transport.

Although the demonstration of a direct effect on contractile activity on amino acid transport appears very exciting and

encouraging to us, the physiological significance of the phenomena
is presently unclear. Studies with the transport of natural amino
acids indicates that not all amino acids respond in similar fashion
to AIB. Exactly analogous situations have been found with insulin,
which promotes transport of AIB but not all other amino acids into
muscle [38]. Specifically, exercise does not affect the transport
of the branched chain or other large hydrophobic neutral amino
acids, which utilize a different transport system than AIB. The
significance of these findings remains to be established.

Experiments are also being pursued in an attempt to determine
how these rapid changes in transport might be related to the increased synthesis of RNA and protein which occur during hypertrophy.
An attractive working hypothesis is that the changes in amino acid
transport constitute the initial event of work-induced growth and
that increased levels of amino acids intracellularly somehow lead
to other biochemical correlates of work-induced growth (Table 1).
There is, for example, appreciable biochemical evidence from studies
of bacterial systems that amino acid supply plays an important role
in the regulation of net RNA synthesis [45] and protein breakdown
[22]. Although our present data are in accord with this hypothesis,
direct experimental support for this view is totally lacking.

V. AMINO ACID CATABOLISM IN MUSCLE

In the course of these studies, certain rather surprising
observations were made which may affect the interpretation of many
previous observations on amino acid transport and incorporation in
muscle. Traditionally it has been believed that skeletal muscle
burns only glucose and fatty acids, and that amino acids in muscle
are used only for protein synthesis. Recently, however, we have
found that skeletal muscle catabolizes certain amino acids at a
rapid rate; in fact, muscle appears to be the major site in the
body for catabolism of leucine, isoleucine, and valine [26,47].

Our interest in the possibility of appreciable amino acid

catabolism in muscle arose during the experiments on the isolated
rat diaphragm, described above. It was noted that the rates of
leucine entrance into the tissue did not correlate with the rate of
accumulation of leucine in the intracellular pools or in proteins.
Experiments were therefore carried out by Mr. Richard Odessey to
test the possibility that leucine might be converted by the diaphragm to carbon dioxide, which would not have been measured by the
earlier procedures. The excised muscle was incubated with [14]C-
labeled amino acids, and at varying times thereafter the amount of
[14]C-recoverable as CO_2, in protein, and in the acid-soluble pool
was measured. As shown in Table 4, a major fraction of the leucine,
isoleucine, valine, alanine, aspartic acid, and glutamic acids
entering the muscle are catabolized to CO_2 and thus did not enter
protein synthesis. For these residues, oxidation occurred at a
comparable or greater rate than incorporation into protein.

For all other amino acids tested, catabolism to CO_2 was not
measurable or in the case of glycine, serine, or proline accounted
for less than 5% of the amino acid taken up by the muscle [26]. At
the present time, we have no idea why skeletal muscle degrades only
these six amino acids, while other tissues, such as liver or kidney
rapidly catabolize those residues not degraded by muscle. In
addition, little is known of the regulation of amino acid catabolism
in muscle. Addition of large amounts of glucose inhibited the
oxidation of leucine only slightly (15%) while other energy sources
for muscle (e.g., pyruvate, β-hydroxbutyrate, or palmitate) had no
effect on this process. Blockage of energy metabolism with cyanide
or iodoacetate markedly reduced [14]CO_2 production from leucine.
However, inhibition of protein synthesis stimulated leucine oxidation. Apparently because of the block in protein synthesis, amino
acids within the muscle switched from an anabolic to a catabolic
pathway.

Previous studies of protein synthesis have completely ignored
the possibility of appreciable amino acid catabolism. We recently
learned that Manchester [37] independently has obtained evidence

TABLE 4

Amino Acid Metabolism by Rat Diaphragm[a]

Amino acid	$^{14}CO_2$ produced[b]	^{14}C incorporated into protein[b]	^{14}C in TCA-soluble fraction[b]
Leucine-1-^{14}C	38%	51%	11%
Isoleucine-1-^{14}C	36%	40%	24%
Valine-1-^{14}C	25%	48%	26%
Alanine-1-^{14}C	57%	16%	37%
Glutamate-U-^{14}C	47%	6%	47%
Aspartate-U-^{14}C	53%	7%	40%

[a]Quarter diaphragms from normal rats were incubated with the radio-active amino acid (0.1 mM) in Krebs-Ringer-bicarbonate buffer for 90 min. The amount of ^{14}C recovered as CO_2, in protein or in TCA-soluble form was then measured and expressed relative to the total amount of cpm recovered. For other amino acids studied, oxidation did not appear to be a major metabolic pathway. No measurable CO_2 production was observed from threonine, lysine, methionine, phenyl-alanine, histidine, tyrosine, and tryptophan, while $^{14}CO_2$ production from glycine, proline, and serine accounted for less than 5% of the radioactivity recovered from the muscle [26].

[b]cpm recovered/total cpm taken up by muscle.

for the same conclusion, although his earlier report obviously has received little attention. These findings affect a number of es-tablished ideas about muscle and body nitrogen metabolism, and even raise the possibility that oxidation of amino acids may be a sig-nificant source of energy for muscle under certain physiological conditions, such as food-deprivation. Recent experiments [26] have shown that the rate of leucine oxidation increased significantly in muscles of food-deprived rats. In such muscles, incorporation of all the amino acids into protein was markedly reduced. However, in the same tissues, catabolism of leucine, isoleucine, or valine to

CO_2 increased three- to fivefold above rates found in muscles from normal rats (Table 5). Starvation significantly increased the total uptake of the branched chain amino acids into the diaphragm, although no change was evident in the radioactivity of the intracellular pools. Interestingly this increased ability of muscle to burn branched chain amino acids in food-deprived rats coincides with increased concentrations of these residues in the blood, and also coincides with a reduced tendency of muscle to burn glucose.

In other physiological states where muscle growth is decreased, such as diabetes or following hypophysectomy, there also appears to be simultaneously increased rates of oxidation of these three amino acids and decreased rates of protein synthesis. Thus it appears possible that the regulation of the amino acid degradation may be intimately related to the control of growth and protein turnover in muscle. Future studies will attempt to clarify which biochemical factors determine whether these six amino acids are degraded by the muscle or incorporated into new proteins.

VI. PROTEIN CATABOLISM IN MUSCLE

In addition to the above studies, we have also been actively engaged in studies of the catabolism of muscle proteins during growth and atrophy. Although it has been realized for almost thirty years [51] that proteins in mammalian cells are continually degraded and replaced, very little is known at present of the mechanisms or control of protein degradation. Our ignornace in this area is in sharp contrast to our sophisticated knowledge about protein synthetic mechanisms. As a result, most studies of mammalian growth have been concerned only with synthetic events and have totally ignored the possibility that changes in protein half-lives can influence organ size. In the past several years, there has been a growing recognition of the importance of protein degradation in controlling intracellular enzyme levels [50]. Our own studies indicate that average rates of protein catabolism in cells change

TABLE 5

Comparison of Amino Acid Metabolism in Diaphragms from Fed and
Fasted Rats[a]

Amino acid	^{14}C recovered in			
	CO_2[b]	TCA-soluble[b]	Protein[b]	Total[b]
Leucine-1-^{14}C	2.7[c]	1.0	0.4[c]	1.3[c]
Isoleucine-1-^{14}C	4.5[c]	1.0	0.3[c]	2.0[c]
Valine-1-^{14}C	4.6[c]	1.0	0.5[c]	1.6[c]

[a]The ratio of cpm recoverable in CO_2, protein, and TCA-soluble form
were compared in diaphragms from fed and fasted rats, following
incubation of quarter diaphragms, as in Table 4. Fasted rats were
deprived of food for 3 days. Although incorporation of all amino
acids decreased 50-70% catabolism and total uptake of the branched
chain amino acids increased. At the same time, the amount of cpm
in the TCA-soluble pools did not differ in the two groups. Statis-
tical analyses were based upon comparisons of the means ± S.E.M.
using the Students T-test [26].

[b]Ratio of fasted/fed.

[c]Significant difference between fasted and fed rats ($p < 0.05$).

under different physiological states in a manner that influences
organ size.

Even under basal conditions, rates of protein breakdown are
not the same in different muscles of the body. For example, in
hypophysectomized rats, we observed [13] several years ago that the
darker (tonic) muscles of the body appeared more active in protein
synthesis than the pale (phasic) ones. Since these observations
were made in nongrowing animals, they must also indicate more rapid
rates of protein breakdown in the dark muscles. More interesting
than these variations in basal rates was the demonstration that
hormones and the level of muscular work can markedly affect these
degradative rates in muscle.

Our initial experiments [16] were undertaken to determine

whether changes in catabolic rates might contribute to compensatory
growth. To compare protein breakdown in hypertrophying and nongrowing muscles, muscle proteins were initially labeled with ^{3}H-
leucine. Several days later tenotomy of the gastrocnemius was
performed and the loss of labeled material was compared in the
contralateral muscles. (Various experimental controls were performed to minimize possible complications caused by the recycling
of radioactive amino acids through proteins.) These experiments
indicated that during hypertrophy, in addition to the increased
protein synthesis, there occurred a significant reduction in catabolism of muscle proteins, especially sarcoplasmic ones. Together
the effects on synthesis and degradation should be additive in
promoting gain in muscle weight. Our previous data on amino acid
incorporation [14] during hypertrophy could not by themselves
account quantitatively or qualitatively for the accumulation of
proteins during hypertrophy; however, consideration of the findings
on protein breakdown helped eliminate these difficulties. Interestingly, reduced protein degradation was specific to work-induced
growth and was not seen when muscle growth was induced by treatment
with pituitary growth hormone, which only affected protein synthesis.

Changes in degradation also appear of fundamental importance
in the process of muscle atrophy. Theoretically, a tissue might
decrease in size in several ways: (a) by decreasing protein synthesis such that protein catabolism becomes dominant, (b) by
increasing degradative processes without changing synthetic rates,
and (c) by some combination of changes in both synthesis and breakdown. All of these possible mechanisms have been suggested to
explain denervation atrophy. To decide between these alternatives,
we have carried out analogous experiments to those described above
and have compared the fate of labeled proteins in contralateral
denervated and control muscles [17]. As shown in Fig. 3 the
denervated muscle consistently contained less of the labeled
material, indicating increased protein degradation. This effect
was especially marked for the myofibrillar proteins, which account

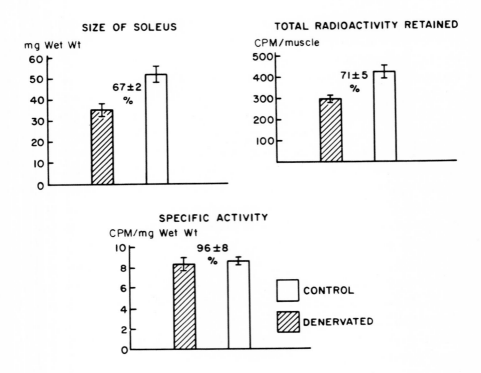

Figure 3. Effects of denervation on labeled proteins in soleus
muscle. Values are the mean ± standard error obtained in seven
animals 10 days after denervation. The findings with the denervated
muscle are also expressed as percentage of contralateral innervated
muscle (i.e., mean % ± S.E.M.). The decrease in muscle weight and
in total counts per minute per muscle following denervation are
significant at $p < 0.01$, and indicate increased protein catabolism
during denervation atrophy. No change in specific activity occurred
($p < 0.1$), suggesting decreased protein synthesis following denervation.

for the selective depletion of these components following denervation. By varying the time after denervation when the animals were
sacrificed it became clear that the magnitude of this apparent
increase in protein catabolism was directly related to the loss of
muscle weight. Addition of hormones which influenced the magnitude
or rate of denervation atrophy did not alter the experimental
results (Fig. 4). Despite the increased loss of labeled proteins,
the specific activity of proteins in the denervated and control
muscles did not differ. This finding indicates a concomitant
decrease in new protein synthesis simultaneous with the heightened
breakdown of "old" components (Fig. 3). Together these two effects
on protein metabolism should have complementary results in decreasing muscle mass.

In related studies, we examined the mechanism of atrophy
induced by large doses of cortisone. This form of muscle atrophy
is an experimental equivalent to that seen in the clinical syndrome,
Cushings Disease, and probably represents an extreme version of the
mobilization of muscle proteins induced in "hard times" by adrenal
steroids [44]. The extent of muscle wasting induced by glucocorticoids depends on the level of muscular work and is most
pronounced in the least active muscles [19]. Like denervation,
cortisone both promoted protein breakdown and inhibited further
protein synthesis. However, the increased protein catabolism
differed in specificity in these two types of atrophy. The glucocorticoids promoted breakdown of myofibrillar and sarcoplasmic
fractions proportionately, while denervation most affected the
myofibrillar proteins. An important area for future research would
be to learn more about the degree of specificity of these changes
in average degradative rates perhaps by following breakdown of
individual proteins under these various physiological conditions.

Although this in vivo approach provided important physiological
information, it was not very useful for studies of the mechanisms

of proteolysis and was not sufficiently sensitive for evaluating
degradative rates over short period of time. Recently Richard
Fulks in our laboratory has developed a new, simple procedure for
studying protein catabolism in isolated rat diaphragm [11]. When
incubated in vitro this muscle shows linear incorporation of amino
acids for several hours, but at the same time shows a linear
release of amino acids into the incubation medium (as evidenced by
the appearance of α-amino nitrogen or tyrosine). This release of
amino acids appears to be a specific physiological process reflect-
ing protein breakdown rather than simply loss of intracellular

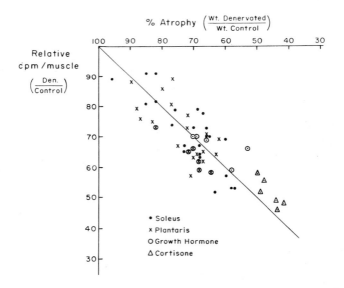

Figure 4. Relationship of the loss of weight to the loss of
labeled protein in denervated muscles (i.e., to the increased
protein breakdown). The animals were sacrificed 5, 10, or 15 days
after nerve section. The solid line represents the best fit to this
data by the method of least squares (slope of -1.0). Also included
are comparisons of contralateral denervated and innervated muscles
in animals treated with ovine growth hormone (1 mg/rat/day) or
cortisone (10 mg/rat/day), both of which influence the absolute
rates of denervation atrophy [18,19].

amino acid pools. Under these in vitro conditions, protein degradation in the muscle exceeded protein synthesis, as would be found in an atrophying tissue in vivo (e.g., in muscles of starving animals).

As shown in Fig. 5, addition of insulin, either in the presence or absence of glucose, inhibited the release of α-amino nitrogen. This effect would be consistent with the well-known anabolic effects of insulin. Additional experiments were carried out to determine whether insulin's action could be accounted for simply by its well-known effect in stimulating protein synthesis, or whether this hormone might also inhibit proteolysis in muscle. By simultaneous measurements of release of tyrosine, the incorporation of ^{14}C-tyrosine into protein, and the intracellular pools within the diaphragm, we found that the inhibition of tyrosine release was several times greater than could be explained by insulin's promotion of tyrosine incorporation [11]. Further evidence for a direct hormonal effect on protein breakdown is shown in Fig. 5. In muscles in which protein synthesis was completely blocked by cycloheximide, insulin was still capable of decreasing amino acid release by decreasing intracellular proteolysis. Recently insulin has been reported to reduce protein breakdown in cardiac muscle [55] and in liver [43]. Of special interest is the finding that the insulin can affect independently three different aspects of protein metabolism: protein synthesis, amino acid transport, and protein breakdown.

It is hoped that this simple experimental system will permit us to study how other physiological factors, such as other hormones or muscular contraction, influence the catabolic process. Other experiments by Mr. Fulks have shown that glucose by itself reduces protein catabolism in muscle. This observation is of interest because of the old observation that carbohydrate in the starving organism spares body nitrogen [44]. These effects of carbohydrate have generally been believed to be mediated by release of pancreatic insulin. Our observations raise the possibility that the supply of glucose in the circulation by itself might directly regulate the net catabolism of muscle proteins. Studies of this possibility are in progress.

Of special interest to us is the cellular mechanisms that might relate the changes in degradative and biosynthetic events (e.g., how insulin and increased work might simultaneously increase protein

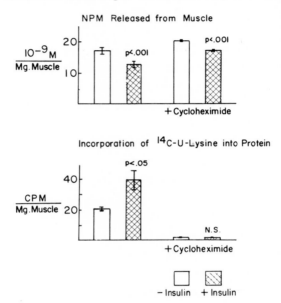

Figure 5. Effects of insulin on protein breakdown and synthesis in isolated rat diaphragm. Diaphragms were removed and preincubated for 1/2 hr in Krebs-Ringer-bicarbonate prior to transfer to fresh medium containing ^{14}C-U-lysine. NPM represents TCA-soluble ninhydrin-positive material (primarily amino acids) produced by the muscle during the subsequent 3 hr. Because intracellular pools did not change significantly, the release of NPM reflects net protein breakdown. Insulin significantly reduced release of amino acids, partially by stimulating amino acid incorporation (^{14}C-U-lysine). Since insulin had similar effects in the presence of cycloheximide, insulin must have an independent inhibitory effect on protein breakdown. Values are the mean ± standard error obtained in seven quarter diaphragms. Standard errors were calculated by the method paired analysis, based on comparisons of different pieces of the same diaphragm.

synthesis and decrease protein breakdown). In an attempt to learn
more about the interrelationships between these different aspects
of protein metabolism, we have also been investigating control of
protein breakdown in growing and nongrowing E. coli. These related
studies [24] have demonstrated that intracellular levels of aminoacyl-
tRNA can influence average rates of proteolysis. In addition to
indicating that the intermediates in protein synthesis can exert
a feedback influence on degradation, these studies strongly suggest
that the control of protein breakdown in cells is coordinated with
the control of ribosome biosynthesis. A similar suggestion has
been put forward for mammalian cells in culture by Hershko and
Tompkins [31]. Whether similar control mechanisms are found in
skeletal muscle appears likely but remains to be demonstrated.

Much further work is essential if we are to understand the
biochemical mechanisms responsible for these alterations in protein
breakdown. Of great importance in evaluating the physiological
importance of these findings on muscle would be better information
on the actual half-lives of muscle proteins. An outstanding question is whether the various contractile proteins turn over at the
same or different rate. This information obviously has important
implications for understanding the assembly of the myofibril, its
dissolution during atrophy, or its maintenance.

The literature about turnover rates in muscle is in a highly
confused and confusing state. For example, it is still frequently
claimed that there is very little turnover of muscle proteins, and
half-lives of the order of weeks or even months are frequently
quoted on the basis of very weak evidence. Valid data on degradative rates are difficult to obtain because of a variety of possible
technical complications. The most serious complication is the
possibility of recycling radioactive amino acids within the tissue,
following a single administration of radioactive precursors to the
organism. This problem is especially serious in studies of skeletal
muscle, in which certain radioactive amino acids (e.g., lysine),
are retained in the intracellular pools for days or even weeks

following a single intravenous injection (unpublished observations).
The very long half-lives or anomalous decay patterns for muscle
proteins [49] probably result from this artifact.

Recent elegant studies by Millward [40] have clearly demonstrated how the use of different radioactive precursors can yield
different apparent half-lives for muscle proteins. These studies
indicate that the actual mean half-life of muscle proteins to be
approximately 10 days. These values are in accord with our own
unpublished findings. This value was obtained using radioactive
precursors, which are rapidly degraded by muscle. It is of interest
in this context that our own studies on breakdown in vivo used
leucine, alanine, or glutamic acid, all of which we now know are
rapidly degraded by muscle.

VII. POSTSCRIPT

This discussion is a report on work in progress. Although
some success has been achieved in defining compensatory growth and
in characterizing this process, we still do not know how changes in
muscle work lead to increased muscle mass. Despite a number of new
findings concerning muscle growth, amino acid metabolism, and
protein turnover, the relationship between these observations and
the control of size is far from clear. Further progress is limited
by our lack of knowledge in several areas of cell physiology.

The work thus far has clearly pointed out a number of important
unanswered questions which we are presently investigating. For
example, the finding that muscular contraction directly influences
amino acid transport, while interesting in itself, actually raises
fundamental questions about the relationship between the control of
transport and control of amino acid utilization. Similarly the
discovery that muscle catabolizes certain amino acids at an appreciable rate has led us to explore the physiological role of this
pathway and its importance in control of protein synthesis and
degradation. Finally, the discovery that changes in rates of

protein breakdown contribute to muscle growth clearly emphasizes
the necessity of greater knowledge about the mechanisms responsible
for protein breakdown, their regulation, and their relationship to
protein synthesis. It is probably naive to anticipate significant
progress in understanding growth or atrophy until such fundamental
problems are first clarified.

ACKNOWLEDGMENTS

These experiments were initiated in the laboratory of Dr. H.
M. Goodman and I am deeply grateful to Dr. Goodman for his advice
and encouragement. I am also indebted to my other collaborators in
this work, Richard Odessey, Richard Fulks, Dr. Charles Jablecki,
and especially Mrs. Susan Martel.

These studies have been supported through grants from the Air
Force Office of Scientific Research, the Berkshire County Chapter of
the Massachusetts Heart Association, and the Muscular Dystrophy
Association of America.

REFERENCES

[1] Alpert, N. (ed.), Cardiac Hypertrophy, Academic, New York, 1971.

[2] Arvill, A., Acta Endocr., 56, suppl. 122 (1967).

[3] Bartsova, D., M. Chvapil, R. Korecky, O. Poupa, K. Rakusan,
Z. Turek, and M. Vizek, J. Physiol., 200, 285 (1969).

[4] Beznak, M., J. Physiol., 116, 74 (1952).

[5] Bigland, R. and R. J. Jehring, J. Physiol., 166, 129 (1952).

[6] Buccino, R. A., E. Harris, and J. F. Spann, Jr., Am. J. Physiol.,
216, 425 (1969).

[7] Christensen, H. W., Biological Transport, Benjamin, New York,
1963.

[8] De La Haba, G., G. W. Cooper, V. Elting, Proc. Natl. Acad. Sci.,
U.S., 56, 1719 (1967).

[9]   Evans, H. M., M. E. Simpson, and C. H. Li, Growth, 12, 15 (1948).

[10] Fanburg, B. L., New Engl. J. Med., 282, 723 (1970).

[11] Fulks, R. and A. L. Goldberg (submitted for publication).

[12] Goldberg, A. L., Am. J. Physiol., 213, 1193 (1967).

[13] Goldberg, A. L., Nature, 216, 1219 (1967).

[14] Goldberg, A. L., J. Cell Biol., 36, 653 (1968).

[15] Goldberg, A. L., Endocrinol., 83, 1071 (1968).

[16] Goldberg, A. L. and H. M. Goodman, J. Physiol., 200, 655 (1969).

[17] Goldberg, A. L. and H. M. Goodman, J. Physiol., 200, 667 (1969).

[18] Goldberg, A. L. and H. M. Goodman, Am. J. Physiol., 216, 1111 (1969).

[19] Goldberg, A. L. and H. M. Goodman, Am. J. Physiol., 216, 1116 (1969).

[20] Goldberg, A. L., J. Biol. Chem., 244, 3217 (1969).

[21] Goldberg, A. L., J. Biol. Chem., 244, 3223 (1969).

[22] Goldberg, A. L., in Ref. [1], p. 39.

[23] Goldberg, A. L., in Ref. [1], p. 301.

[24] Goldberg, A. L., Proc. Natl. Acad. Sci., U.S., 68, 362 (1971).

[25] Goldberg, A. L., C. Jablecki, and S. Martel (submitted for publication).

[26] Goldberg, A. L. and R. Odessey (submitted for publication).

[27] Goss, R. J., Adaptive Growth, Logos Press, London, 1964.

[28] Goss, R. J., Science, 153, 1615 (1966).

[29] Grove, D., R. Zak, and K. G. Nair, Clin. Res., 16, 231 (1968).

[30] Hamosh, M., M. Lesch, J. Baron, and S. Kaufman, Science, 157, 935 (1967).

[31] Hershko, A. and G. Tompkins, J. Biol. Chem., 246, 710 (1971).

[32] Knobil, E. and J. Hotchkiss, Ann. Rev. Physiol., 26, 4774 (1964).

[33] Knobil, E., Physiologist, 9, 25 (1966).

[34] Korner, A., Progr. Biophys., 17, 61 (1967).

[35] Kostyo, J. C. and A. F. Redmond, Endocrinology, 79, 531 (1966).

[36] Lesch, M., R. Gorlin, and E. H. Sonnenblick, Circulation
Research, XXVII, 445 (1970).

[37] Manchester, K. L., Biochim. Biophys. Acta, 100, 295 (1965).

[38] Manchester, K. L., Mammalian Protein Metabolism (H. N. Munro,
ed.) Vol. IV, Academic, New York, 1970, p. 229.

[39] Manchester, K. L., Biochem. J., 117, 457 (1970).

[40] Millward, D. J., Clin. Sci., 39, 577 (1970).

[41] Morkin, E. and T. P. Ashford, Am. J. Physiol., 215, 1409 (1968).

[42] Moroz, L. A., Circulation Res., 21, 449 (1967).

[43] Mortimore, G. E. and C. E. Mondon, J. Biol. Chem., 245, 2375
(1970).

[44] Munro, H. N. and J. B. Allison, Mammalian Protein Metabolism,
Vol. 1, Academic, New York, 1964.

[45] Neirlich, D. P., Proc. Natl. Acad. Sci., U.S.A., 60, 1345
(1968).

[46] Norman, T. D., Prog. Cardiovasc. Dis., 4, 439 (1962).

[47] Odessey, R. and A. L. Goldberg (submitted for publication).

[48] Riggs, T., in Action of Hormones on Molecular Processes
(G. Litwack and D. Kritchersky, eds.), Wiley, New York, 1964.

[49] Schapiro, G., J. C. Dreyfus, and J. Kruh, in Effects of Use
and Disease on Neuromuscular Function (E. Gutmann and P. Hnik, eds.)
American Elsevier, New York, 1963, p. 407.

[50] Schimke, R. T., in Mammalian Protein Metabolism (H. N. Munro,
ed.), Vol. IV, Academic, New York, 1970.

[51] Schoenheimer, R., Dynamic State of Body Constituents, Harvard
University Press, Cambridge, Mass., 1942.

[52]   Schreiber, S. S., M. Oratz, and M. A. Rothschild, <u>Am. J.</u>
<u>Physiol</u>., 213, 1552 (1967).

[53]   Smith, P. E., <u>Am. J. Anat</u>., 45, 205 (1930).

[54]   Wool, I. G., <u>Federation Proc</u>., 24, 1060 (1965).

[55]   Morgan, H., (1971) (personal communication).

CHAPTER 6

EFFECTS OF ADRENERGIC NEUROTRANSMITTERS ON THE UTERUS*

Jean M. Marshall

Section of Neurosciences
Division of Biological and Medical Sciences
Brown University
Providence, Rhode Island

I. INTRODUCTION

Over 60 years ago Sir Henry Dale [10] drew attention to the remarkable fact, noted almost simultaneously by Cushney [9], that electrical stimulation of the hypogastric (adrenergic) nerves caused the uterus of the pregnant cat to contract and of the nonpregnant to relax. Dale coined the term "pregnancy reversal" for this phenomenon. These original observations have been repeatedly

*The author's investigations were supported by National Institutes of Health Grant HE-10187.

confirmed and amplified. We now know that the response of the
myometrium to adrenergic nerve stimulation or to adrenergic amines,
epinephrine and norepinephrine, may be either excitatory or inhibitory depending upon the hormonal status of the individual and on
the species of animal being studied. The implication of the ovarian
hormones in "pregnancy reversal" was first suggested by Gustavson
and Van Dyke [19]. They found that the uterus of a spayed cat
injected with an extract of corpus luteum contracted in response to
hypogastric nerve stimulation while that of a cat injected with
follicular fluid relaxed. In the rabbit pregnancy reversal is just
the opposite from that in the cat; the pregnant rabbit uterus
relaxes and the nonpregnant contracts upon hypogastric nerve stimulation [33,35]. Treatment of an immature or spayed rabbit with
estrogen favors adrenergic stimulation of the uterus whereas estrogen followed by progesterone favors inhibition [30].

In all species regardless of hormonal or gravid status of the
individual the hypogastric nerves are predominantly adrenergic and
release the same transmitter, norepinephrine or a mixture of
norepinephrine and epinephrine. Therefore, the nature of the
uterine response to adrenergic amines lies within the myometrial
cell itself. Whether it is excitatory or inhibitory is determined
by the combination of the neurotransmitter with one or more specific
molecular sites on or within the muscle cell. These sites are
usually called "receptors," a convenient term for concealing our
ignornace of the chemical and morphological nature of these reactive
regions. The classification of adrenoceptors into alpha and beta
groups is now generally accepted [3]. All uteri contain both alpha
(excitatory) and beta (inhibitory) receptors [29] and the response
to adrenergic amines depends, among other things, on the relative
dominance of one or the other of these two adrenoceptors. In the
myometrium the relative dominance of either receptor is apparently
under hormonal control and may vary from one species to another.

The present discussion focuses on the effects of adrenergic
amines and adrenergic nerve stimulation on the myometrium first at

the cellular and tissue levels and then at the organ level. At the
cellular level the possible modes of action of the neurotransmitters
will be considered while at the organ level some functional aspects
of adrenergic nerve stimulation on uterine motility will be noted.
In the latter connection the morphological characteristics of the
adrenergic innervation of the uterus will be reviewed briefly.

II. EFFECTS AT THE CELLULAR AND TISSUE LEVELS

As mentioned previously, the response of the myometrial cell
to the adrenergic amines is determined by the relative dominance of
the alpha or beta receptive sites. If these sites are located on
the excitable membrane surrounding the muscle cell, then changes in
membrane potential should accompany the action of the adrenergic
amines. That this is the case is shown in Fig. 1. Cells stimulated
by these amines are usually depolarized while those inhibited are
usually hyperpolarized. In addition to illustrating the membrane
potential changes accompanying the excitatory and inhibitory actions
of epinephrine, the tracings in Fig. 1 also serve to show the
species differences in the uterine response to this amine. The
hormonal influence on the response of the rat uterus to norepinephrine is shown in Fig. 2. In the rat, epinephrine relaxes the uterus
irrespective of its hormonal state [8,24], but norepinephrine
relaxes uteri from ovariectomized rats or from spayed animals
treated with a combination of estrogen and progesterone [12].
Blockade of the inhibitory (beta) receptors by propranolol prevents
this inhibition and unmasks the stimulatory (alpha) effects of this
amine (Fig. 2). Uterine segments from spayed rats treated only
with estrogen are stimulated by norepinephrine and this action is
reversed in the presence of an alpha blockade (phentolamine)
(Fig. 2) [12].

These actions are also observed in uteri from rats at different
stages of pregnancy. Epinephrine relaxes the uterus throughout
pregnancy and during parturition while norepinephrine generally

relaxes during early and midpregnancy but stimulates at term.   The
inhibitory effects of epinephrine are accompanied by a hyperpolarization of the cell membrane, but in the presence of a beta blocker
this action is converted to a stimulation and depolarization (top
tracings, Fig. 3).   Thus, although epinephrine is capable of activating both alpha and beta receptors, the latter effects are dominant
and stimulation appears only in the presence of a beta blocker.
The stimulatory effects of norepinephrine at term are reversed by

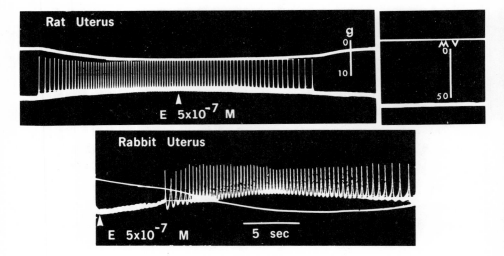

    Figure 1.   Comparison of the effects of epinephrine (E) on the
membrane potentials, lower trace, and isometric tension, upper trace,
with increase in tension downward deflection, recorded from isolated
strips of rat and rabbit uteri at term pregnancy.   Upper frames:
effect of E applied during spontaneous contraction of muscle.   In
about 15 sec, action potentials disappear and muscle begins to relax.
Five minutes later (interval between frames) membrane potential has
increased (hyperpolarized) to about 60 mV and muscle is completely
relaxed.   Lower frame: effect of E when applied to quiescent muscle.
Membrane slowly depolarizes, small potential fluctuations eventually
reach threshold and evoke action potentials which, in turn, increase
contractile force.   Both rate and force of spontaneous contractions
increase (not shown on this record).   (From Marshall, 1967, by permission of Federation of American Societies for Experimental Biology.)

an alpha blocker (lower frames, Fig. 3). These effects at the end
of gestation resemble those in the estrogen injected animals (cf.
Fig. 2) and imply that the alpha receptors may have become more
sensitive to norepinephrine at this time. This change in sensitivity in the pregnant animals could be related to the rising level
of circulating estrogen at the terminal stage of pregnancy in the
rat [6]. In other words, estrogen might be sensitizing the alpha
receptors to norepinephrine.

Since the changes in membrane potentials accompanying the
stimulatory and inhibitory effects of the adrenergic amines are

RAT UTERUS

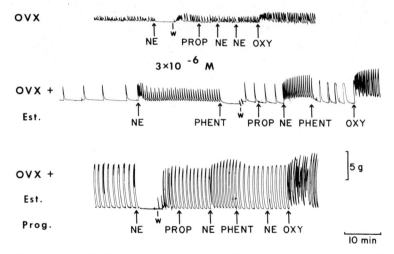

Figure 2. Influence of estrogen (Est) and progesterone (Prog)
on the response of the isolated rat uterus to norepinephrine (NE)
3×10^{-6} M. OVX, uterus from ovariectomized rat; OVX + Est, uterus
from ovariectomized rat treated with 50 µg estradiol benzoate 48 hr
before experiment; OVX + Est + Prog, uterus from ovariectomized rat
primed with 50 µg estradiol benzoate followed 48 hr later by 25 µg
estradiol + 3 mg progesterone daily for 3 days. PROP, propranolol
HCl, 10^{-6} g/ml; PHENT, phentolamine HCl, 2×10^{-6} g/ml; OXY,
oxytocin 2 mU/ml; W, wash. (From Marshall, 1967, by permission of
Federation of American Societies for Experimental Biology.)

prevented or reversed by appropriate adrenoceptor blocking agents,
it is conceivable that the adrenoceptors may influence the mechanisms responsible for the potential gradient across the cell membrane.
The changes in membrane potential caused by epinephrine and norepinephrine may reflect alterations in the ionic permeability or
ionic concentration gradients across the cell membrane, as a result
of a combination of the amines with the adrenoceptor sites.

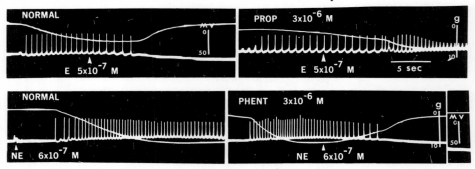

Figure 3.   Comparison of effects of E and NE on uterine segments from two parturient rats, before (normal) and after adrenoceptor blocking agents.   Top trace isometric tension, bottom trace
transmembrane potential.   Upper left frame: E given during spontaneous contraction, action potentials abolished, membrane hyperpolarizes to around 60 mV, muscle relaxes.   Contractions cease for
about 10 min (not shown here).   Upper right frame: same muscle
after propranolol HCl (PROP), E given near end of spontaneous
contraction, after about 15 sec action potential frequency increases,
membrane depolarizes, contractile force increases.   Lower left
frame: second muscle, NE given during silent period between spontaneous contractions, action potentials initiated, membrane depolarizes, spike frequency increases, muscle contracts.   Lower right
frame: same muscle in presence of phentolamine HCl (PHENT).   NE
given during spontaneous contraction now abolishes spikes, muscle
relaxes, membrane hyperpolarizes.   Break between frames signifies
5 min.   Contraction inhibited for about 10 min (not shown here).
(From Marshall, 1967, by permission of Federation of American
Societies for Experimental Biology.)

Therefore attempts have been made to explain some of the effects of epinephrine and norepinephrine in terms of their actions on ionic movements and ionic permeabilities in the individual smooth muscle cells [11,17]. The individual ionic permeabilities for sodium, potassium, chloride, and calcium have not been determined directly in the myometrium because of the technical difficulties encountered when working with these small cells surrounded by extensive connective tissue networks and associated with secretory epithelium and nervous elements. Measurements of ionic movements and distributions using tracer technique are complicated and sometimes give equivocal or contradictory results due to the complexities inherent in kinetic analysis of a multi-compartment system such as the myometrium [11,18]. Deductions about changes in ionic permeabilities are often made indirectly by observing changes in the membrane potential of the myometrial cells in solutions of different ionic compositions. Ideally this type of experiment should also include measurements of tissue electrolytes and of ion fluxes, but the combination of electrophysiology and tracer analysis has been done on the myometrium only in a few instances [5,11,21]. The following account summarizes results from both types of experiments.

The resting potential of the myometrial cell is thought to be predominantly a potassium diffusion potential although both sodium and chloride ions probably also contribute [21,22]. A significant permeability to both sodium and chloride ions as well as to potassium may account for the relatively low resting potential, around 50 mV, in the myometrium as compared with the potassium equilibrium potential, around 80 mV. The excitatory effects of epinephrine and norepinephrine could result from a nonspecific increase in permeability to various ions including sodium, potassium, calcium, and chloride [1, 17,26]. Consequently the membrane potential would move to a level somewhere between the equilibrium potentials for these ions and in so doing would decrease. Action potentials would be generated when the potential reached firing threshold. Evidence for a nonspecific

increase in ionic permeability is the finding that the excitatory
effects of epinephrine and norepinephrine on the rabbit uterus
(cf. Fig. 1) are dependent upon normal concentrations of sodium,
potassium, and calcium ions in the external medium [26]. Inhibition
with its accompanying hyperpolarization could result from one or a
combination of several of the following: (a) a selective increase
in potassium permeability bringing the membrane closer to the
potassium equilibrium potential, (b) a selective decrease in sodium
permeability or in chloride permeability allowing potassium to
dominate the potential and bringing the potential level nearer the
potassium equilibrium potential, (c) stimulation of an electrogenic
sodium pump, e.g., an active pumping of sodium ions out of the cell
unaccompanied by anions, and (d) an increase in the fixation of
calcium within the membrane. Stimulation of such a pump as (c)
would hyperpolarize the membrane as a result of the net efflux of
positive charge. With respect to (d), in many excitable tissues,
including smooth muscle, calcium and sodium are thought to compete
for the same site in their movement through the membrane [40]. It
has been suggested, but not proved, that the relatively high resting
permeability for sodium in smooth muscle might be due to a poor
fixation of calcium in the membrane which would then allow more
sites for sodium to exchange across the membrane [22]. An increase
in the binding of calcium would reduce the number of sodium sites
diminishing the sodium permeability and hyperpolarizing the membrane.
Calcium is known to be a membrane stabilizer in a variety of excitable cells [40]. Let us now consider the evidence for or against
each of these possibilities.

A. Increase in Potassium Permeability

If epinephrine selectively increases the membrane permeability
to potassium then the amount of hyperpolarization caused by epinephrine should be related to the potassium gradient across the membrane.
Figure 4 compares the membrane potentials from quiescent myometrial
cells (i.e., potentials measured between spontaneous contractions

of the muscle) with those from muscles whose spontaneous activity
has been abolished by epinephrine at different concentrations of
extracellular potassium.   In these experiments, potassium chloride
replaced equivalent amounts of sodium chloride so that the extracellular chloride concentration was unchanged.   The black dots in
Fig. 4 designate the potassium equilibrium potential (E_k) calculated
according to the Nernst equation.   If the membrane were permeable
exclusively to potassium ions then as the external potassium is
reduced the membrane potential should fall along the line connecting
the black dots.   From Fig. 4 we can see that at all concentrations

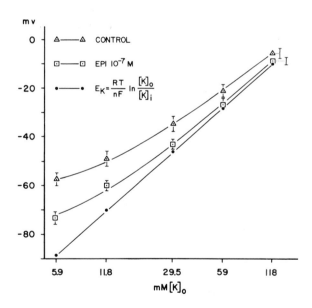

Figure 4.   Effect of varying external potassium concentration,
$[K]_o$, on the membrane potentials from quiescent myometrial cells
normally and after inhibition by epinephrine (EPI).   Rat uterus at
term pregnancy.   Solid dots indicate values for potassium equilibrium
(E_k) calculated from data of Casteels and Kuriyama, 1965.   (From
Marshall, 1968, by permission of Federation of American Societies
for Experimental Biology.)

epinephrine brings the membrane potential closer to E_k. Furthermore the hyperpolarization produced by epinephrine is diminished in parallel with E_k. These electrophysiological findings support the idea that epinephrine selectively increases the potassium permeability of the cell membrane.

B. Decrease in Sodium Permeability

Evidence for a decrease in sodium permeability (P_{Na}) during the inhibitory action of beta adrenergic amines has recently been given by Daniel and colleagues [11] who showed that the sodium content of rat uterine muscle decreased during exposure to isoproterenol (a beta adrenergic agent). They proposed that this decrease was the result of a reduction in sodium permeability of the cell membrane resulting in a diminished influx of sodium into the cell. There is no direct electrophysiological evidence for a decrease in P_{Na} since sodium conductance has not been measured during inhibition produced by adrenergic amines. However, the finding that hyperpolarization persists in the absence of extracellular sodium ions would seem to argue against a decrease in P_{Na} being a major component of the inhibition [26].

C. Stimulation of an Electrogenic Sodium Pump

Although the presence of an electrogenic sodium pump has recently been demonstrated in rat myometrium [41], the contribution of this pump to the membrane potential normally or during drug action is questionable. In some excitable tissues the pump is dependent on the external presence of potassium and can be short-circuited by small anions such as chloride, nitrate, and iodide [31]. Therefore, if the hyperpolarization accompanying the actions of adrenergic amines is mediated by stimulation of an electrogenic sodium pump these actions should be abolished in potassium-free or chloride-free media. However, this is not the case [44]. Furthermore, replacement of extracellular sodium by sucrose or Tris chloride does not prevent the hyperpolarizing actions of epinephrine

on the rat uterus [26]. Ouabain (10^{-3}M) or cooling the muscle to 10°C
reduces but does not abolish hyperpolarization [44]. On balance,
then, the inhibitory actions of the adrenergic amines do not appear
to be associated with stimulation of electrogenic sodium pumping in
the uterus but they might have an energy-requiring component.

D. Increased Calcium Binding in the Membrane

Although there is no direct evidence that beta adrenergic
amines increase ^{45}Ca uptake or Ca binding [11], the hyperpolarization accompanying inhibitory actions of these amines disappears in
a calcium free medium [26]. This latter finding is in partial
agreement with suggestion (d), since in the absence of external
calcium the influence of the amines on calcium binding would be
minimal. Another corollary of the calcium binding hypothesis,
however, is that an increase in calcium in the external medium
should accentuate the inhibitory effects of epinephrine and related
amines. Just the opposite is true for the rat myometrium where
elevation of extracellular calcium antagonizes the effects of
epinephrine [13]. Calcium ions, however, may be involved at some
other site along the pathway between the excitable membrane and the
contractile elements since calcium can antagonize the inhibitory
effects of the adrenergic amines.

The inhibitory actions of the adrenergic amines are not mediated exclusively by alterations in the electrical gradients across
the myometrial cell since they are still present when the cell
membrane has been depolarized by isotonic potassium solution to a
level where all electrical activity is abolished [13,36,37]. The
effects on the depolarized muscle are antagonized by the appropriate
adrenoceptor blocking agents, indicating a site or sites of action
at some point beyond the electrically polarized membrane. The
adrenergic amines are potent metabolic stimulants, and it has been
suggested that the relaxant action of the beta adrenergic amines is
mediated by an increase in the level of cyclic AMP within the

myometrial cells [11,32,42]. Recent studies show that the time
course for the increase in cyclic AMP parallels that for membrane
hyperpolarization and relaxation in rat myometrium [44].

III. EFFECTS AT THE ORGAN LEVEL

The functional significance of "pregnancy reversal" or of the
hormonal regulation of myometrial response to the adrenergic trans-
mitters is not obvious. One of the reasons for this is that our
knowledge and appreciation of the anatomy and physiology of the
adrenergic innervation of the myometrium has been limited. Recently,
however, the techniques of histochemistry, biochemistry, and electron
microscopy have been utilized in studies of the adrenergic influences
on uterine activity. The results of these investigations have added
significantly to our knowledge of the adrenergic innervation of the
uterus. The following discussion attempts to review briefly some
of this information. A more extensive account appears elsewhere
[28].

A. Anatomy and Morphology of Adrenergic Innervation

Recent evidence indicates that the adrenergic innervation of
the uterus possesses at least three unique properties all of which
may be important to our understanding of the role of these nerves
in uterine function. These properties are: (a) The adrenergic
neurons innervating the myometrium arise from pelvic ganglia located
in or near the uterus. Therefore the short, post-ganglionic fibers
coming from these ganglia do not degenerate when the pre-ganglionic
nerves are cut or when the spinal cord is sectioned. Some degree
of adrenergic nerve activity is thus retained at the local,
peripheral level even though the extrinsic nerves are severed.
(b) The density of adrenergic innervation varies from species to
species and in different portions of the myometrium in any one
species. (c) The transmitter content of the nerves and density
of uterine innervation are altered during pregnancy and after
the administration of ovarian hormones. This suggests that the

amount and distribution of the adrenergic transmitter liberated
during nerve activity depends upon the gravid or hormonal state of
the individual. Each of these properties will now be considered in
more detail.

1. *Extrinsic and Intrinsic Innervation*

      The pelvic adrenergic nerves originate in the lumbar region of
the spinal cord and exit in the white rami communicantes to the
fourth and fifth lumbar ganglia of the sympathetic chain from which
they run to the inferior mesenteric ganglia. The hypogastric nerves
arise from the inferior mesenteric ganglia and course down either
side of the midline, where they are easily visualized and isolated
for experimental studies (see below). Many years ago Langley [23]
showed that the hypogastric nerves contain both pre- and post-
ganglionic fibers, the pre-ganglionic synapsing in the ganglia of
the pelvic plexus or within the pelvic viscera, the post-ganglionic
synapsing in the vertebral ganglia. This anatomic arrangement is
shown diagramatically in Fig. 7.

      The adrenergic nerves going to the myometrium, the so-called
uterine nerves, are the short post-ganglionic fibers whose cell
bodies are in the pelvic plexus or within the uterine walls. The
precise nature of the relation between the terminal adrenergic nerve
network and the myometrial cells was never discernable with conventional histologic and microscopic techniques. Recently histochemical fluorescence microscopy has been particularly useful in
the identification and characterization of the terminal adrenergic
nerve network within various organs including the myometrium.
The basis of this technique, which was developed over ten
years ago by Eränko and by Falck and associates (see [14] for
details), is the formation of an intense fluorescence by monoamines
in freeze-dried tissues exposed to formaldehyde vapor. The epinephrine and norepinephrine within the adrenergic fiber can be detected
when the tissues are sectioned and examined under a fluorescence
microscope. The number and arrangement of terminal adrenergic
neurons within the myometrium can be visualized in this manner.

Along with this advance in histochemistry came the development of
sensitive chemical methods for the tissue analyses of adrenergic
amines. The richness of innervation could then be checked and
correlated with parallel chemical analyses of tissue amines.

Some of the characteristic features of the intrinsic adrenergic
nerves in the myometrium as revealed by histochemical fluorescence
microscopy are shown in Fig. 5 taken from the work of Dr. N. -O.
Sjöberg [39]. One of the striking features of the terminal nerve
network is the presence of bead-like varicosities along the entire
length of the fine nerve terminals. Chemical analyses of the
epinephrine and norepinephrine content of the tissue when correlated
with the size and distribution of the varicosities provide indirect
evidence that the adrenergic neurotransmitter is located at these
regions [15], in that the more intense the fluorescence the more
concentrated the transmitter. The varicosities are distributed all
along the terminal axons, forming a sort of "synapse en passage."
In this manner the transmitter is released along the terminal
axon when an action potential sweeps over the nerve. One axon
can thereby influence many muscle cells. This arrangement is in
contrast to that in the somatic motor nerve where the transmitter
is localized and released only from discreet nerve terminals at the
motor end plate. Another prominent difference between the neuroeffector mechanism in visceral and skeletal muscle has recently
been revealed by electron microscopy which permits the visualization
of the relation between the nerve fiber and smooth muscle cells.
The neuroeffector junctions in smooth muscle are regions where small
axons partially or completely devoid of their Schwann cell sheaths
lie within 200 or 300 Å of the smooth muscle sarcolemma. Unlike
the motor end plate of skeletal muscle, the nerve and smooth muscle
cells are not specialized at their points of apposition. The regions
of nerve axons lying in close association with the smooth muscle
cells are believed to be the bead-like varicosities, since they
usually contain vesicles, some of which are electron dense, suggesting the presence of the adrenergic amines.

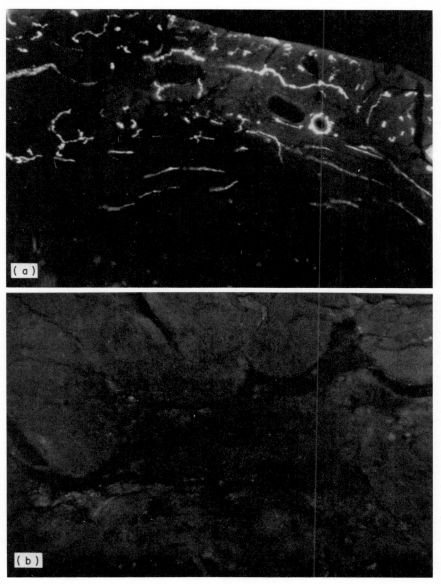

Figure 5. Adrenergic innervation of myometrium from the guinea
pig as visualized by fluorescence microscopy. (a) Uterus from nonpregnant animal. Endometrium (below), inner circular layer (middle),
and outer circular layer (middle), and outer longitudinal layer
(above). Moderate number of fluorescent, varicose nerve terminals
running mainly in direction of smooth muscle cells 125 X. (b) Uterine horn, end of pregnancy. The musculature and blood vessels are
devoid of fluorescent norepinephrine-containing nerves 200 X. (From
Sjöberg, 1968, by permission of P.A. Norstedt and Sons.)

133

Various other important characteristics might be mentioned
about the intrinsic adrenergic nerves in the myometrium [28].
First the predominant amine found within the terminal network is
norepinephrine with little or no epinephrine present. Second, the
stores of norepinephrine are not depleted when the extrinsic (hypogastric or sacral) nerves are cut, providing further evidence for
the idea that the cell bodies of these nerves reside in or near the
uterus. Third, the stores of transmitter within the nerves are
depleted more slowly by reserpine than similar stores in other
adrenergic neurons in the heart and spleen. Fourth, the rate of
turnover of amines, that is synthesis and release, is much slower
in the granules of adrenergic nerve terminals from the uterus than
from other organs. Taken together these findings emphasize that
adrenergic denervation of the uterus cannot be accomplished by
removal of the vertebral ganglia, section of the spinal cord, or
the hypogastric nerve, or by the administration of reserpine and
other transmitter depleting agents.

2. *Density of Innervation*

The adrenergic innervation of the cat uterus is richer than
that of the rabbit or the guinea pig. These findings correlate
well with the four- to fivefold higher norepinephrine content of
the cat uterus. In contrast to the cat, rabbit, and guinea pig,
the adrenergic nerve supply to the rat uterus is limited almost
exclusively to the vascular system [28].

The only species examined thus far which shows regional variations in the density of uterine adrenergic innervation is the
human where the cervix has the highest innervation density as
compared with the corpus or fundus. It is possible that this distribution of innervation allows a more precise neural control of
the cervix which may be of special importance during delivery and
immediately postpartum [28].

3. *Alteration of Innervation Density during Pregnancy and After*
    Administration of Ovarian Hormones

It is possible to estimate changes in the density and transmitter content of the adrenergic nerves within the uterus during
pregnancy or after hormonal treatment if the number and arrangement
of the nerves and their transmitter content are correlated with the
mass of muscle tissue.

Such a study has been made in the guinea pig by Dr. Sjöberg at
the University of Lund [39]. He selected animals at different times
in pregnancy or during estrus and after injections of estrogen or
progesterone. There was little or no change in transmitter content
or nerve distribution in the myometrium during the first part of
gestation and the innervation patterns characteristic of this time
in pregnancy resembled those in estrous animals or in spayed animals
given estrogen. After midpregnancy, however, there was a dramatic
decrease both in distribution and in content of transmitter so that
by the end of gestation hardly any nerves were visible within the
myometrium [Fig. 5(b)]. The marked reduction in adrenergic elements
at the end of pregnancy could be mimicked by the administration of
progesterone or by the selection of animals in metestrus.

Thus both the innervation density as well as transmitter
content of the adrenergic neurons within the myometrium varies with
the gravid or hormonal status of the individual.

B. Effects of Adrenergic Nerve Stimulation

The question now arises, what is the physiological significance
of these morphological and biochemical findings? It is not possible
to give a definitive answer to this important query since much less
is known about the functional properties of the uterine adrenergic
nerves than about their morphology and chemistry.

Several preparations of isolated hypogastric nerve-uterine
muscle preparations have been devised [20,43] so that the effects

of electrical stimulation of the adrenergic nerves on uterine
activity can be studied in an isolated organ bath for many hours.
The rabbit and guinea pig have proven especially useful in such
studies because their hypogastric nerves are easily visualized and
can be isolated along with the uterine horns. The results of one
such study in the rabbit showed that electrical stimulation of the
hypogastric nerves caused the uterus from an estrogen-treated rabbit
to contract and that from an estrogen-progesterone-treated rabbit
to relax [30]. These effects were accompanied by alterations in the
membrane potential of the individual myometrial cell. As one might
expect in an estrogen-dominated animal, nerve stimulation caused a
depolarization of the membrane leading to spike discharge while in
a progesterone-dominated one the spontaneous spike discharge was
abolished and the membrane was hyperpolarized (Fig. 6). These
actions were mimicked by the addition of norepinephrine to the
bathing medium and were prevented by alpha and by beta receptor
blocking agents, respectively. The reversal of response from
excitation to inhibition in the progesterone-dominated animal
occurred before any measureable changes in tissue catecholamines
were detected [30].

In order to elicit a clear-cut response of the uterus (inhibition or excitation), it was necessary to stimulate the hypogastric
nerve at frequencies between 20 and 50 pulses per second, a range
considerably above the physiological discharge frequency for most
autonomic efferents including the uterine nerves [4,16]. Hence the
physiological significance of the uterine response under these circumstances is questionable.

Recently Dr. Meinhard Rüsse and I have attempted to study the
effects of hypogastric nerve stimulation on uterine motility in the
anesthetized guinea pig [34]. Our primary objective was to investigate the effects of hypogastric nerve stimulation at frequencies
below 15 pulses per second on the uterine motility in vivo in both
nonpregnant and pregnant animals in order to parallel the histochemical and biochemical work of Sjöberg [39]. Guinea pigs
were anesthetized with urethane, the abdominal cavity opened, and

the hypogastric nerve cut just below the inferior mesenteric plexus.
The distal ends of the nerves were threaded through platinum loop
electrodes embedded in Lucite.   Intrauterine pressure, which served
as an index of uterine motility, was monitored with a saline-filled
catheter inserted into one uterine horn.   After the catheter and
the electrodes were in place, the abdomen was sutured shut.   When
the uterus became quiescent, usually in about 30 min, the experiment
was begun.   A diagram of the experimental arrangement appears in
Fig. 7 which also outlines the innervation pathway.

In the estrous animal stimulation of the hypogastric nerve at
frequencies above 15 pulses per second for periods of 2 to 5 min
usually caused a contraction of the uterus.   However, after three

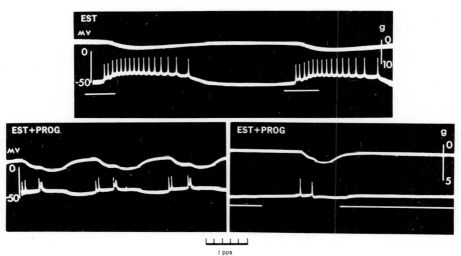

Figure 6.   Effect of hypogastric nerve stimulation on transmembrane potentials (lower traces) and isometric tension (upper
traces, increase in tension downward) of an isolated strip of rabbit
myometrium.   White lines beneath transmembrane potential trace
indicate periods of nerve stimulation.   Stimulus parameters: 20
pulses/sec, supramaximal voltage, 5 msec pulse duration.   Est,
estrogen-dominated uterus; Est + Prog, estrogen-primed progesterone-
dominated uterus.   (From Marshall, 1969, by permission J & A
Churchill, Ltd.)

or four consecutive periods of stimulation the contractions become
weaker and eventually disappeared. If the abdomen was opened at
this time the uterine blood vessels were found gorged with blood
and had a bluish appearance in contrast to their normally pink or
red color in the unstimulated uterus. Apparently vascular changes
had occurred during nerve stimulation which in turn reduced uterine
contractility. At stimulation frequencies below 15 pulses per
second there was no consistent contraction of the uterus and no
observable change in uterine vasculature. Although no contractions
occurred during nerve stimulation at these lower, presumably more
physiological frequencies, it seemed possible that the nerves might
still be exerting a more subtle action on the myometrium. For
example the uterine sensitivity to circulating hormones might be

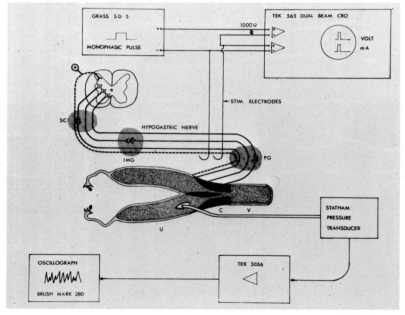

Figure 7. Schematic drawing of experimental preparation of
hypogastric-nerve-uterus preparation in the guinea pig. SC, sympathetic chain; IMG, inferior mesenteric ganglion; PG, pelvic ganglion;
U, uterus; C, cervix (dark area); V, vagina; dotted line, afferent
nerve pathway. (From Rüsse and Marshall, 1970, by permission of
Academic Press.)

altered during adrenergic nerve stimulation as suggested by Abrahams et al. [2]. To test this possibility we examined the effects of various intravenous doses of oxytocin on uterine motility before and during hypogastric nerve stimulation in guinea pigs in estrus. The stimulation frequency in these experiments, 4 to 6 pulses per second, had no effect per se on uterine motility. When nerve stimulation was combined with oxytocin injection, however, the uterine response to the hormone was potentiated as shown in Fig. 8 which illustrates a typical experiment. The potentiation was prevented by phentolamine (5 x 10^{-6} g/kg body weight) and was slightly accentuated by propranolol (5 x 10^{-6} g/kg body weight). It was also prevented by the ganglion blocking agent hexamethonium (5 mg/kg body weight). Thus the potentiation was mediated by post-ganglionic adrenergic axons whose transmitter, norepinephrine, was activating predominantly alpha (excitatory) adrenoceptors.

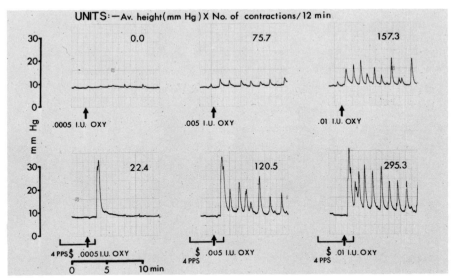

Figure 8.   Increase in uterine sensitivity to oxytocin accompanying hypogastric nerve stimulation in the guinea pig in estrus. OXY, oxytocin, i.v.; ⅋ PPS, stimulation of nerve with rectangular pulses of 1.5 msec duration 1.0 mA intensity; numbers at top of each frame represent units of uterine motility. Time interval between frames is 20 min. (From Rüsse and Marshall, 1970, by permission of Academic Press.)

Our experiments do not provide an explanation for the mechanisms underlying the potentiating effects of nerve stimulation on
the uterine response to oxytocin. However, one could predict on
the basis of other evidence [26] that norepinephrine might decrease
the threshold of the myometrial cells to stimulatory agents such as
oxytocin. It has been suggested [7] that oxytocin will not trigger
activity in myometrial cells which are unable to discharge propagated action potentials. Therefore, it is conceivable that the
adrenergic neurotransmitter norepinephrine enhances the effects of
oxytocin by improving the overall excitability of the myometrium.
As a result muscle cells previously insensitive to oxytocin because
of their "borderline" excitability now respond to this hormone.

When these experiments were repeated on guinea pigs during the
early weeks of pregnancy the results were similar to those just
mentioned for the estrous animals. At the end of pregnancy and
immediately postpartum, however, no significant effect of nerve
stimulation on the oxytocin response was observed even in the
presence of propranolol indicating that the lack of response was
not due to a relative increase in beta (inhibitory) adrenoceptors
at term pregnancy (Fig. 9). These results correlate well with the
histochemical and biochemical data of Sjöberg which indicated a
marked reduction in innervation density and in norepinephrine content of the uterus at the end of pregnancy. Apparently the innervation does not keep pace with the great increase in muscle size and
weight during the last days of pregnancy. Under these circumstances
the amount of transmitter liberated per gram of myometrium during
nerve stimulation is not sufficient to enhance the overall uterine
excitability, and therefore the response to oxytocin is not potentiated.

Although the physiological significance of these observations
is not immediately obvious, it is conceivable that the overall
excitability of the myometrium might be influenced and modulated by
tonic activity of the adrenergic nerves within the muscle. This
does not imply, however, that such activity is the only regulatory
influence since it is well known that other factors such as the

ovarian and neurophyophysial hormones and stretch are also of great
importance in modulating uterine excitability. Nevertheless, to
this list we might now add a neural influence. In the guinea pig
the adrenergic component would tend to augment uterine excitability
by an effect on the alpha adrenoceptors. However, at the end of
pregnancy when the other influences, namely estrogen, stretch, and
oxytocin are probably already maximal [38], the need for an additional neural component may be minimal. Of course this suggestion
needs to be varified by comparative investigations on a variety of
species utilizing physiological, morphological, and biochemical
techniques.

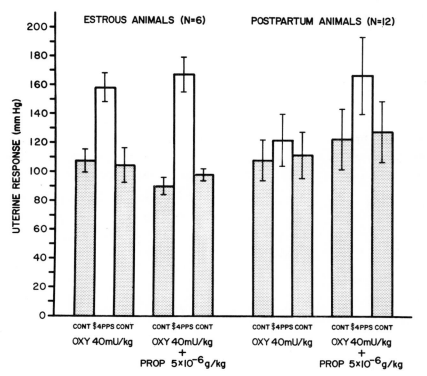

Figure 9. Effects of hypogastric nerve stimulation on uterine
motility in the anesthetized guinea pig. ⌀ 4PPS is nerve stimulation with rectangular pulses, 5 msec duration, 1.0 mA intensity for
5 min. OXY, oxytocin, i.v.; PROP, propranolol HCl, i.v. Height of
each column equals mean value, ± S.E.M. (Rüsse and Marshall
unpublished observations.)

REFERENCES

[1] Abe, Y., "The hormonal control and the effects of drugs and
ions on the electrical and mechanical activity of the uterus," in
Smooth Muscle (E. Bülbring, A. Brading, A. W. Jones, and T. Tomita,
eds), Williams and Wilkins, Baltimore, 1970, p. 396-417.

[2] Abrahams, V. C., E. P. Langworthy, and G. W. Theobald, "Potentials evoked in the hypothalamus and cerebral cortex by electrical
stimulation of the uterus," Nature, 203, 654-656 (1964).

[3] Ahlquist, R. P., "A study of the adrenotropic receptors,"
Am. J. Physiol., 153, 586-599 (1948).

[4] Bower, E. A., "The activity of post-ganglionic sympathetic
nerves to the uterus of the rabbit," J. Physiol. London, 183,
748-767 (1966).

[5] Casteels, R. and H. Kuriyama, "Membrane potential and ionic
content in pregnant and non-pregnant rat myometrium," J. Physiol.
London, 177, 263-287 (1965).

[6] Catchpole, H. R., "Endocrine mechanisms during pregnancy," in
Reproduction in Domestic Animals, Vol. 1 (H. H. Cole and P. T. Cupps,
eds.), Academic Press, New York, 1959, p. 469-508.

[7] Csapo, A., "Defense mechanism of pregnancy," in Progesterone
and the Defense Mechanism of Pregnancy, Ciba Foundation Study Group
No. 9 (G. E. W. Wolstenholm and M. P. Cameron, eds.), Churchill,
London, 1961, p. 3-31.

[8] Csapo, A. I. and H. A. Kuriyama, "Effects of ions and drugs on
cell membrane activity and tension in the postpartum rat myometrium,"
J. Physiol. London, 165, 575-592 (1963).

[9] Cushney, A. R., "On the movements of the uterus," J. Physiol.
London, 35, 1-19 (1906).

[10] Dale, H. H., "On some physiological actions of ergot,"
J. Physiol. London, 34, 163-206 (1906).

[11] Daniel, E. E., D. M. Paton, G. S. Taylor and B. J. Hodgson, "Adrenergic receptors for catecholamine effects on tissue electrolytes," Federation Proc., 29, 1410-1425 (1970).

[12] Diamond, J. and T. M. Brody, "Hormonal alterations of the response of the rat uterus to catecholamines," Life Sci., 5, 2187-2193 (1966).

[13] Diamond, J. and J. M. Marshall, "A comparison of the effects of various smooth muscle relaxants on the electrical and mechanical activity of the rat uterus," J. Pharmacol. Exptl. Therap., 168, 21-30 (1969).

[14] Eränko, O., "Histochemistry of nervous tissues: Catecholamines and cholinesterases," Ann. Rev. Pharmacol., 7, 203-222 (1967).

[15] Falck, B., "Observations on the possibilities of the cellular localization of monoamines by a fluorescence method," Acta Physiol. Scand., 197, 1-25 (1962).

[16] Folkow, B., "Impulse frequency in sympathetic vasomotor fibres correlated to the release and elimination of the transmitter," Acta Physiol. Scand., 25, 49-76 (1952).

[17] Gillespie, J. S., "The mode of action of catecholamines on smooth muscle," Mem. Soc. Endocrinol., 14, 155-170 (1966).

[18] Goodford, P. J., "Ionic interactions in smooth muscle," in Smooth Muscle (E. Bülbring, A. F. Brading, A. Jones, and T. Tomita, eds.), Williams and Wilkins, Baltimore, 1970, p. 100-121.

[19] Gustavson, R. G. and H. B. Van Dyke, "Further observations on the pregnancy response of the uterus of the cat," J. Pharmacol. Exp. Therap., 41, 139-146 (1931).

[20] Isaac, P. R., J. N. Pennefather, and D. G. Silva, "The ovarian and hypogastric innervation of the guinea pig uterus," Europ. J. Pharmacol., 5, 384-390 (1969).

[21] Kao, C. Y., "Ionic basis of electrical activity in smooth

144   JEAN M. MARSHALL

muscle," in Cellular Biology of the Uterus (R. M. Wynn, ed.),
Appleton, Century, Crofts, New York, 1967, p. 386-448.

[22] Kuriyama, H., "Ionic basis of smooth muscle action potentials,"
in Handbook of Physiology, Section 6 Alimentary Canal (C. F. Code
and W. Heidel, eds.), Am. Physiol. Soc., Washington, D. C., 1968,
Vol. IV, p. 1767-1791.

[23] Langley, J. N. and H. K. Anderson, "The constituents of the
hypogastric nerves," J. Physiol. London, 17, 177-191 (1894).

[24] Marshall, J. M., "Effects of estrogen and progesterone on
single uterine muscle fibers in the rat," Am. J. Physiol., 197,
936-942 (1959).

[25] Marshall, J. M., "Comparative aspects of the pharmacology of
smooth muscle," Federation Proc., 26, 1104-1110 (1967).

[26] Marshall, J. M., "Relation between the ionic environment and
the action of drugs on the myometrium," Federation Proc., 27,
115-119 (1968).

[27] Marshall, J. M., "The effects of ovarian hormones on the
uterine response to adrenergic nerve stimulation and to adrenergic
amines," in Progesterone: Its Regulatory Effect on the Myometrium,
Ciba Foundation Study Group No. 34 (G. E. W. Wolstenholme and J.
Knight, eds.), Churchill, London, 1969, p. 89-105.

[28] Marshall, J. M., "Adrenergic innervation of the female
reproductive tract: Anatomy, physiology and pharmacology," Ergebn.
Physiol., 62, 7-67 (1970).

[29] Miller, J. W., "Adrenergic receptors in the myometrium,"
Ann. N. Y. Acad. Sci., 139, 788-798 (1967).

[30] Miller, M. D. and J. M. Marshall, "Uterine response to nerve
stimulation; relation to hormonal status and catecholamines,"
Am. J. Physiol., 209, 859-865 (1965).

[31] Rang, H. P. and J. M. Ritchie, "On the electrogenic sodium

pump in mammalian non-myelinated nerve fibres and its activation by various external cations." J. Physiol. London, 196, 183-221 (1968).

[32] Robison, G. A., R. W. Butcher, and E. W. Sutherland, "On the relation of hormone receptors to adenyl cyclase," in Fundamental Concepts in Drug-Receptor Interactions, Academic Press, New York, 1970, p. 59-91.

[33] Rudolph, L. and A. C. Ivy, "The physiology of the uterus in labor," Am. J. Obstet. Gynec., 19, 317-335 (1930).

[34] Rüsse, M. W. and J. M. Marshall, "Uterine response to adrenergic nerve stimulation in the guinea pig," Biol. Reproduction, 3, 13-22 (1970).

[35] Sauer, J., C. E. Jett-Jackson, and S. R. M. Reynolds, "Reactivity of the uterus to pre-sacral nerve stimulation and to epinephrine, pituitrin and pilocarpine administration during certain sexual states in the anesthetized rabbit," Am. J. Physiol., 111, 250-256 (1935).

[36] Schild, H. O., "Calcium and the effect of drugs on depolarized smooth muscle," in Pharmacology of Smooth Muscle (E. Bülbring, ed.), Pergamon Press, Oxford, 1964, p. 95-104.

[37] Schild, H. O., "The action of isoprenaline in the depolarized rat uterus," Brit. J. Pharmacol., 31, 578-592 (1967).

[38] Schofield, B. M., "Parturition," in Advances in Reproductive Biology (A. McLaren, ed.), Academic Press, New York, 1968, Vol. 3, p. 9-33.

[39] Sjöberg, N. -O., "Considerations on the cause of disappearance of the adrenergic transmitter in uterine nerves during pregnancy," Acta Physiol. Scand., 72, 510-517 (1968).

[40] Somlyo, A. P. and A. V. Somlyo, "Vascular smooth muscle. I. Normal structure, pathology, biochemistry and biophysics," Pharmacol. Rev., 20, 197-272 (1968).

[41] Taylor, G. S., D. M. Paton, and E. E. Daniel, "Characteristics of electrogenic sodium pumping in the rat myometrium," J. Gen. Physiol., 56, 360-375 (1970).

[42] Triner, L., N. I. A. Overweg, and G. G. Nahas, "Cyclic 3', 5'-AMP and uterine contractility," Nature, 225, 282-283 (1970).

[43] Varagíc, V., "An isolated hypogastric-nerve-uterus preparation, with observations on the hypogastric transmitter," J. Physiol. London, 132, 92-99 (1956).

[44] Kroeger, E., and J. M. Marshall, "Beta-adrenergic effects on rat uterus," The Pharmacologist, 13, 686 (1971).

CHAPTER 7

THE STRUCTURE AND FUNCTION OF INTRAFUSAL MUSCLE FIBERS

R. S. Smith and W. K. Ovalle

The Neurophysiology Laboratory
Department of Surgery
University of Alberta
Edmonton, Alberta
Canada

I.   INTRODUCTION

The inclusion in this volume of a review devoted entirely to the properties of intrafusal muscle fibers could easily appear to be a disproportionate representation. Intrafusal muscle fibers make up

only some fractional percentage of the bulk of any skeletal muscle
and consequently they do not contribute in any significant way to
the force which the muscle can exert on its tendons. Special consideration of intrafusal fibers is justified, however, when one
takes note of the role these fibers play in the control of skeletal
muscle. These muscle fibers are the vertebrate representation of
the wider class of receptor muscles: those muscles whose primary
function is not to move the animal but to control the discharge of
a sense organ. In the vertebrates the most intensively investigated organ of this type is the muscle spindle. The muscle spindle
lies in parallel with the majority of muscle fibers in a muscle.
Thus, it is not in a favorable position to sense contraction of the
extrafusal muscle; rather, it is thought that the most important
"inputs" to a muscle spindle are change of length of the muscle and
motor excitation of the intrafusal muscle fibers themselves. Both
of these classes of inputs combine in the intrafusal muscle fibers
to produce deformation of sensory endings which are situated on
these muscle fibers (see Fig. 22). The importance of the resulting
afferent signals in the control of skeletal muscle is not considered
here; this aspect of the function of muscle spindles has been
treated extensively elsewhere [74,122,123]. We consider recent
advances in the understanding of the intrinsic properties of intrafusal muscle fibers. This consideration will be only partly
directed toward their importance in the motor control system. The
intrafusal muscle fiber, as will be seen, also poses many other
problems whose solution will lead to a greater general understanding
of the biology of muscle.

II. THE STRUCTURE OF INTRAFUSAL MUSCLE FIBERS IN THE ADULT ANIMAL

Most of the work on the morphology of intrafusal muscle fibers
has been conducted with the light microscope. It is only recently
that electron microscopic techniques have been used. Several
reviews and symposia [5,8,53,73,74,122] and a bibliographic survey
[61] have recently been devoted in part to this subject. The

fusimotor innervation of intrafusal muscle fibers has also been reviewed [11]. The present discussion is directed mainly to our present knowledge of the general morphology and ultrastructure of intrafusal muscle fibers. The structure of fusimotor nerve endings and sensory nerve terminals are mentioned only insofar as they modify the structure of the intrafusal fibers.

A. Light Microscopic Appearance

The presence of intrafusal muscle fibers of large and small diameter in mammalian spindles has long been known [49,50,153]. Recently two structural types of intrafusal fibers have been described in the mammal [9,22,23,24,39,40]: the nuclear bag muscle fiber and the nuclear chain muscle fiber. A bimodal distribution of diameter and length and the characteristic nuclear arrangement of different mammalian intrafusal fibers, for example, has been reported by several workers [29,60,164,168]. It should be noted that most light microscopists have failed to find a similar structural dichotomy of intrafusal fiber types in lower vertebrates (i.e., in amphibians, reptiles, and birds). Gray [75] suggested that amphibian intrafusal fibers were of two types on the basis of their selective fusimotor innervation (see also Smith [154,155]). Barker and Cope [13], in addition, demonstrated two populations of intrafusal fibers in the frog on the basis of fiber diameter.

Mammalian nuclear bag fibers are generally greater in length and larger in diameter than nuclear chain fibers. Equatorial regions of nuclear bag fibers are conspicuous and contain many tightly packed vesicular nuclei and peripherally placed myofilaments. Equatorial regions of nuclear chain fibers, in contrast, are less conspicuous and contain a single central row of elongated nuclei [Fig. 1(a)]. Juxtaequatorial or "myotube" regions of both fibers are characterized by a single row of centrally placed nuclei surrounded by a peripheral layer of myofibrils. The polar regions, in turn, are characterized by peripherally placed nuclei and groups of myofilaments which occupy almost the entire cross section of each muscle fiber [Fig. 1(b)].

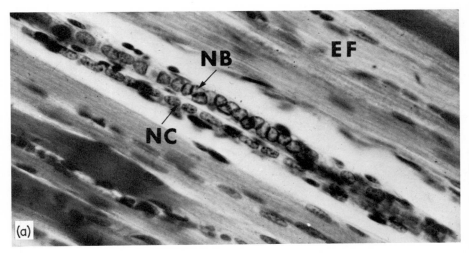

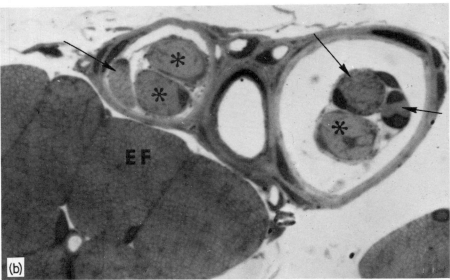

Figure 1. Light micrographs. (a) Longitudinal section of a
muscle spindle in the equatorial region. Note the large diameter
nuclear bag fiber (NB) with closely packed vesicular nuclei. The
small diameter nuclear chain fiber (NC), in contrast, contains
elongated nuclei which run in series in the center of the muscle
fiber. Surrounding extrafusal muscle fibers (EF) are indicated.
Stained with hematoxylin and eosin. X 590. (b) Transverse section
of two muscle spindles. The spindle on the left is sectioned
through the polar zone, while the spindle on the right is sectioned

The relative number of nuclear bag and nuclear chain fibers per muscle spindle varies considerably according to the type of muscle and the particular species studied [74]. Nuclear bag fibers usually extend well beyond the spindle capsule, inserting into the perimysium adjacent to the extrafusal muscle fibers [30]. Nuclear chain fibers, in contrast, usually terminate intracapsularly by inserting on adjacent nuclear bag fibers [24] or directly on the capsule wall [30].

The existence of solely two types of mammalian intrafusal fibers has been questioned by Barker and Gidumal [14]. These workers have, in addition, described an "intermediate" intrafusal fiber which closely resembles the nuclear bag fiber. It, however, possesses a smaller aggregation of nuclei in the bag region in addition to a smaller myotube diameter.

A variation in the myofibrillar pattern in mammalian intrafusal fibers has been reported at the light microscope level [14,24]. Boyd maintained that nuclear bag fibers were exclusively "fibrillar" while nuclear chain fibers were exclusively "areal" (see Gray [76] for classification). Barker et al. [14], on the other hand, held that all intrafusal fibers were "fibrillar" for most of their polar length while the "areal" pattern was characteristic of juxtaequatorial regions. In view of this, Matthews [122] stated in his review that myofibrillar variation in intrafusal fibers was not a reliable distinguishing criterion at that time. Myofibrillar pattern differences between nuclear bag and nuclear chain fibers,

through the myotube zone. Nuclear bag fibers (asterisks) and nuclear chain fibers (arrows) are indicated. Numerous blood vessels are closely associated with the capsules of both spindles. Note the neighboring extrafusal muscle fibers (EF). Stained with Azure II - 1% borax. X 1500. NOTE: Figures 1-4, 6-9 and 11-15 are sections taken through muscle spindles from adult rat lumbrical muscles.

however, have been reported at the ultrastructural level and are discussed below.

B. Ultrastructural Appearance

1. *Mammalian Intrafusal Fibers*

Early studies with the electron microscope [35,77,128] failed to resolve any structural differences between individual intrafusal muscle fibers in the mammal. Merrillees [128], for example, merely described various components of the muscle spindle in the rat lumbrical muscle. He noted, in particular, that myofibrils of intrafusal fibers were generally more tightly packed and less conspicuous than those of the surrounding extrafusal fibers. He also noted that triads of the sarcotubular system [i.e., the T-system and sarcoplasmic reticulum (SR)] in the intrafusal fibers were less numerous than triads in the extrafusal fibers. He suggested that "the intrafusal triad does not extend nearly so far around the circumference of its myofibril as does the extrafusal triad." Cheng and Breinin [35] similarly observed that the T-system and SR were poorly developed in intrafusal fibers of the monkey. They, in addition, noted the more frequent occurrence of mitochondria in the intrafusal fibers than in the surrounding extrafusal fibers.

The first ultrastructural study to distinguish clearly between the mammalian nuclear bag and nuclear chain muscle fiber was that of Landon [109] in the rat. Other workers have since extended some of Landon's original observations in different mammals [1,17,41-43, 59,83,136,137,146].

Landon noted that mitochondria were generally larger and longer in the nuclear chain fiber than in the nuclear bag fiber. Others have, in addition, noted that mitochondria are considerably more abundant in the nuclear chain fiber [1,42,43,136] (Figs. 2-5). In contrast to most extrafusal fibers in which mitochondria tend to be oriented transversely on both sides of the Z band, mitochondria of intrafusal fibers are characteristically oriented in a longitudinal direction, running parallel to the myofibrils and often spanning more than one sarcomere [109] (Fig. 4).

Landon [109] originally noted that sarcomere lengths of nuclear
chain fibers in the rat were 20% less than those of comparably

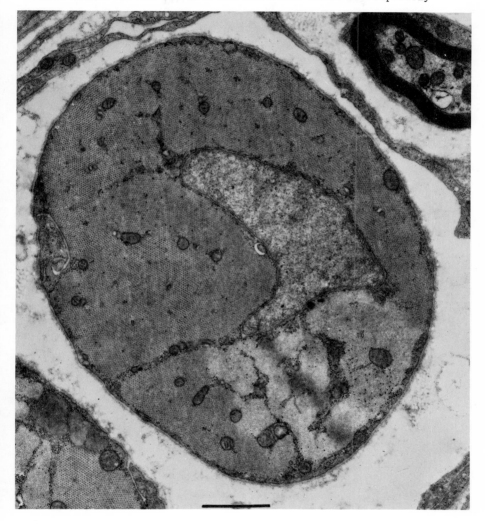

Figure 2. Electron micrograph. Transverse section of a nuclear bag fiber in the polar region. The myofilaments are tightly
aggregated into a continuous bundle giving rise to a "felderstruktur"
appearance. Mitochondria in this fiber are small and sparse.
Stained with uranyl acetate and lead citrate. X 15,870.

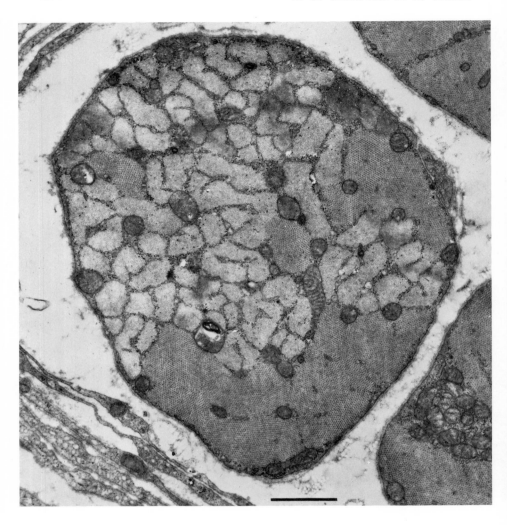

Figure 3. Electron micrograph. Transverse section of a nuclear chain fiber in the polar region of the same muscle spindle as
in Fig. 2. Most of the myofilaments are well delineated into discrete myofibrils and exhibit a "fibrillenstruktur" appearance.
Mitochondria in this fiber are larger and more numerous than those
in the nuclear bag fiber. Stained with uranyl acetate and lead
citrate. X 15,870.

stretched nuclear bag fibers. On the other hand, a worker from
this laboratory (Ovalle, personal observation) has failed to find
any consistent differences in sarcomere length between the two fiber
types in the same animal. It is possible that any apparent differences in sarcomere length may merely be due to different states of
contraction of the two muscle fibers at the time of fixation.

Another criterion which serves to distinguish between the nuclear bag and the nuclear chain fiber is Z band structure. Nuclear
bag fibers generally possess thicker and more conspicuous Z bands
than nuclear chain fibers [42,146] (Figs. 4 and 5).

The myofibrillar pattern of the two intrafusal fiber types in
the mammal is also known to differ, especially in polar regions
[109]. Myofibrils of the nuclear bag fiber are usually more tightly
packed and less discrete giving a "slow-felderstruktur" appearance
in transverse section (Fig. 2). Nuclear chain fibers, in contrast,
possess comparatively well-delineated myofibrils with abundant
interfibrillar sarcoplasm giving a "fast-fibrillenstruktur" appearance (Fig. 3). Such a difference in the geometrical arrangement of
the myofibrils [107] may be related to the observed differences in
the contractile activity of the two intrafusal fiber types [25,26,
156].

The appearance of the M band in a muscle fiber is a structural
criterion which other workers have used to distinguish between
vertebrate "twitch" and "slow" muscle fibers [85,138]. Most "twitch"
fibers, for example, possess M bands while M bands in "slow" fibers
are generally absent. In an analagous fashion most workers [1,17,
42,59,109,146] have described the presence of a well-defined M band
in the nuclear chain fiber (Fig. 6) and the absence of an M band in
the nuclear bag fiber. Upon cursory inspection, particularly at
low magnification (Fig. 4), M bands of a nuclear bag fiber in the
rat are inconspicuous and appear to be absent or barely visible.
However, closer scrutiny of this fiber at higher magnification
(Fig. 7) reveals the presence of an M band, although it is ill-

Figure 4. Electron micrograph. Longitudinal section taken through the polar region of a nuclear chain fiber (top) and a nuclear bag fiber (bottom). The nuclear bag fiber contains thicker Z bands and less prominent M bands than does the nuclear chain fiber. Mitochondria are smaller and less abundant in the nuclear bag fiber. The nuclear chain fiber contains extreme terminal

defined and atypical in appearance [136]. Similar atypical or less
dense M bands have been described in nuclear bag fibers of the cat
[90,150].

At present there is general agreement that a direct correlation
exists in other skeletal muscles between the amount of sarcotubular
system in a given muscle cell and its speed of contraction. Hubbard
and Hess [90] and Adal [1] originally contended that both the nuclear bag and the nuclear chain fiber in the cat contained only
"aberrant" T-system elements. Other workers have more recently
stated that T-tubules and triads are more numerous in nuclear chain
fibers than in nuclear bag fibers [42,43,59,136,146]. In both
intrafusal fiber types, dyads as well as triads are present; they
may be oriented longitudinally as well as transversely and they
may be located at all levels of the sarcomere [136]. This is in
contrast to most adult extrafusal fibers in which triads are
usually oriented transversely and are located solely at the A-I
junctions of each sarcomere. Nuclear chain fibers usually
contain one triad, and often two for every sarcomere (Fig. 8),
while nuclear bag fibers commonly exhibit only one triad for
every two, three, or four sarcomeres (Fig. 9) [136]. The
occurence of even more complex junctional couplings such as
quatrads, pentads, and septads (Figs. 5 and 8) also characterizes
the nuclear chain fiber. Such complex couplings between T-tubules
and SR cisternae are virtually absent in the nuclear bag fiber
(Fig. 5) [136].

The SR is, in addition, considerably more abundant in the
nuclear chain fiber [1,42,43,136]. The appearance of extreme terminal dilitations of the SR, for example, is characteristic of the
nuclear chain fiber, while such dilitations are uncommon in the
nuclear bag fiber (Fig. 4) [136].

dilitations of the sarcoplasmic reticulum (SR). They are usually
filled with a dense granular material and may form intimate contacts
with mitochondria (see right portion of fiber) and with T-tubules.
Stained with uranyl acetate and lead citrate. X 13,750.

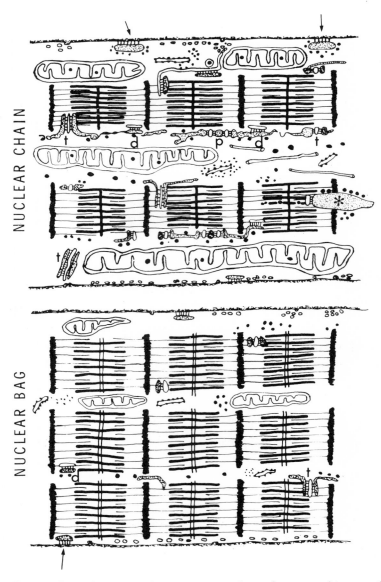

Figure 5. Diagrammatic representation of a mammalian nuclear
bag and nuclear chain muscle fiber. The nuclear chain fiber contains a more abundant and more highly organized sarcotubular system.
SR cisternae in this fiber often exhibit extreme terminal dilitations (asterisk). In addition, a variety of complex junctional
couplings such as dyads (d), triads (t), pentads (p), and peripheral

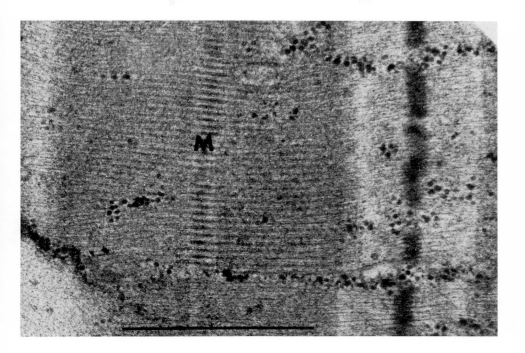

Figure 6.   Electron micrograph.   Transverse section of a portion of a nuclear chain fiber. A typical and conspicuous M band (M) is present in the center of the A band of the sarcomere. Stained with uranyl acetate and lead citrate. X 50,500.

couplings with the sarcolemma (arrows) are abundant in the nuclear chain fiber. In contrast, the SR and T-system of the nuclear bag fiber is poorly developed. Z bands in the nuclear bag fiber appear thicker than those in the nuclear chain fiber. M bands are prominent and appear fairly typical in the nuclear chain fiber, while they are less conspicuous (or absent) in the nuclear bag fiber. Mitochondria are larger and more numerous in the nuclear chain fiber, where they are oriented parallel to the longitudinal axis and often span several sarcomeres. The presence of other organelles and inclusions is more conspicuous in the interfibrillar sarcoplasm of the nuclear chain fiber than it is in the nuclear bag fiber.

From the foregoing it is evident that mammalian nuclear chain
fibers contain a more highly complex and more abundant sarcotubular
system than the nuclear bag fiber. This difference, in addition to
some of the other structural criteria mentioned above, suggests
that nuclear chain fibers more closely resemble extrafusal "twitch"
fibers while nuclear bag fibers more closely resemble extrafusal
"slow" fibers [109].

Equatorial regions of mammalian intrafusal fibers contain a
central core of sarcoplasm filled with a variety of organelles,
inclusions, and nuclei. Each central sarcoplasmic core is, in turn,
surrounded by an attenuated peripheral rim of myofilaments (Fig. 10)
[83,109,128,137]. This is in contrast to the polar regions in

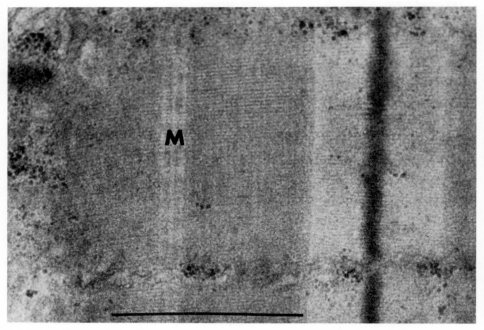

Figure 7. Electron micrograph. Transverse section of a comparable portion of a nuclear bag fiber. An inconspicuous and
atypical M band is present in the center of the A band of the
sarcomere. The M band (M) in this fiber is composed of two parallel
thin densities situated in the center of a poorly defined pseudo-H
zone. Stained with uranyl acetate and lead citrate. X 50,500.

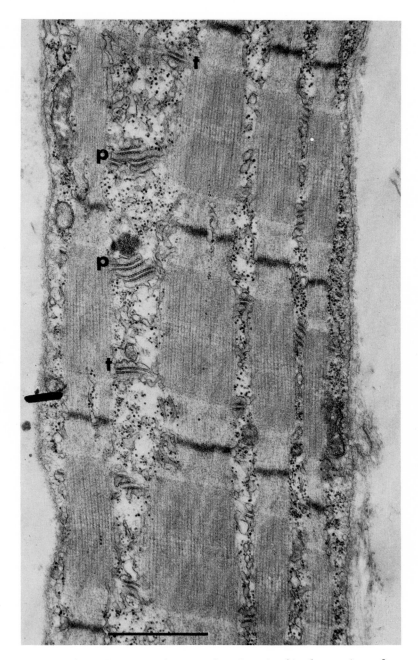

Figure 8. Electron micrograph. Longitudinal section of a
nuclear chain fiber. The sarcotubular system is well developed and
consists of a meshwork of interconnected cisternae passing uninter-
ruptedly from one sarcomere to another. Note the relative abundance
of junctional couplings at the A-I junctions of each sarcomere.
Two triads (t) and two pentads (p) are indicated. Stained with
uranyl acetate and lead citrate. X 26,240.

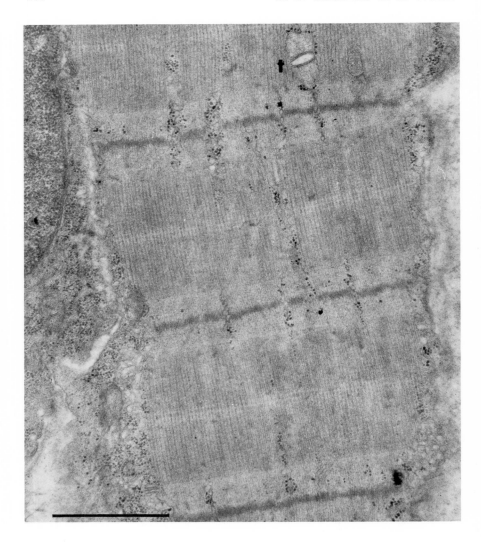

Figure 9. Electron micrograph. Longitudinal section of a
nuclear bag fiber from the same muscle spindle as in Fig. 7. Note
the relative paucity of sarcotubular elements in this fiber. Only
one triad (t) is present in the entire field. Stained with uranyl
acetate and lead citrate. X 30,240.

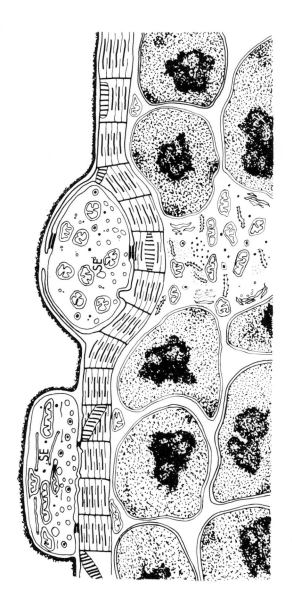

Figure 10. Diagrammatic representation of the equatorial (sensory) region of a mammalian nuclear bag muscle fiber. This region contains many closely packed vesicular nuclei and a central core of sarcoplasm rich in organelles and inclusions. An attenuated peripheral shell of myofilaments is located directly under the sarcolemma. Striated bodies, termed leptomeric organelles, are especially prevalent in this region of the muscle fiber. Some are seen attaching directly to the sarcolemma and others are seen in close association to sensory nerve endings (SE). Two sensory nerve endings (SE) are seen directly under the basal lamina of the muscle fiber. Intimate membrane contacts and close interdigitations between sensory nerve terminal and intrafusal muscle fiber are common.

which almost the entire muscle fiber is occupied by contractile
myofilaments [128]. The process by which the myofilaments decrease
in number from polar to equatorial regions is at present unknown.

Leptomeric organelles, first termed "microladders" by Katz
[102] in the frog, are also present in mammalian intrafusal fibers
[42,77,109,146]. Their precise function is at present unknown.
However, their close association and frequent attachment to the
sarcolemma of the muscle cell directly under a sensory nerve ending
[42] suggests that they may serve to couple the myofibril to the
sarcolemma in the sensory region (Fig. 10). Leptomeres may also be
seen originating at one Z band and terminating at the Z band of an
adjacent sarcomere (Fig. 11). Whether the leptomere is in series
with the myofibril or in parallel with the adjacent sarcomere has
not yet been determined.

Sensory nerve terminals ending on intrafusal muscle fibers are
devoid of a Schwann cell covering. There is characteristically an
intercellular gap, 150-350 Å wide, between sensory ending and muscle
with no intervening basal lamina [1,42,83,109,126,128]. The sarcoplasm of the muscle fiber immediately beneath a sensory ending is
usually occupied by closely packed myofilaments [128]. Figure 12
illustrates two distinct modes of termination of a sensory nerve
ending on an intrafusal muscle fiber in the rat. One sensory nerve
terminal partially invaginates into the surface of the muscle fiber
in such a manner that extensions or "lips" [109] of sarcoplasm
almost cover the bulging nerve terminal. Another sensory terminal
exhibits a flattened or crescent shape and lies directly on the
outer surface of the muscle cell. Close interdigitations and
intimate membrane contacts of the zonula adhaerens variety [126]
frequently occur between sensory nerve terminal and intrafusal
muscle fiber (Figs. 10, 13). Sensory nerve terminals are

characterized by an accumulation of mitochondria, microtubules, neurofilaments, and a variety of clear and dense-core vesicles [1, 42,83,126,128]. The functional significance of such vesicles within sensory nerve terminals is at present unknown.

Figure 11. Electron micrograph. Longitudinal section of a nuclear chain fiber showing two leptomeric organelles (L) spanning different sarcomeres. Note the attachment of both leptomeres to the Z bands (Z) of these sarcomeres (arrows). Stained with uranyl acetate and lead citrate. X 19,400.

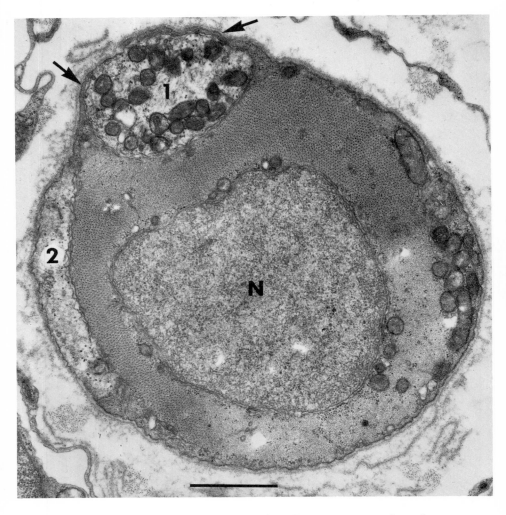

Figure 12. Electron micrograph. Transverse section of a
nuclear chain fiber in the equatorial zone. A centrally located
nucleus (N) is surrounded by a peripheral shell of myofilaments.
One sensory nerve ending (1) invaginates into a depression of the
muscle fiber. Thin extensions of sarcoplasm (arrows) almost enclose
the bulging nerve terminal. Note the abundance of mitochondria
within the sensory ending. Another sensory nerve ending (2) merely
terminates on the surface of the muscle fiber with no extensions of
sarcoplasm formed over it. The basal lamina of the muscle fiber
completely encloses both sensory nerve terminals. Stained with
uranyl acetate and lead citrate. X 22,800.

166

Several workers have described the presence of periodic areas of intimate membrane contact between adjacent nuclear chain fibers in cat [1,41,42] and in rat [126] muscle spindles. Adal [1] designated such areas of convergence as "pairing" of nuclear chain fibers, while Corvaja and co-workers [41,42] characterized them as a zonula adhaerens type of intercellular junction. Such specialized areas of contact are characterized by an intercellular gap of 200 Å or less, no intervening basal lamina, and a dense thickening of the adjacent muscle membranes. Although the precise function of these cell-to-cell junctions is still unknown, it has been suggested that they may represent specialized sites of mechanical attachment between the apposing intrafusal fiber membranes [41], or it is possible that the fibers are electrically coupled via these junctions.

Aside from the two intrafusal fiber types mentioned above, Barker's group [12,16] has recently described another type of intrafusal fiber in the rabbit which exhibits features intermediate to those of the nuclear bag and the nuclear chain fiber. Such "intermediate" fibers possess ultrastructural characteristics similar to those of nuclear chain fibers including the presence of an M band. They are, however, longer than nuclear chain fibers and possess conspicuous nuclear bag (equatorial) regions.

It is now generally agreed that mammalian intrafusal fibers are innervated by at least two kinds of fusimotor nerve endings, "gamma-plates" and "gamma-trails" [11,15,24,27]. Differences in morphology and location of these two forms of motor ending were first detected by cholinesterase staining methods [36,37,84,125]. "Gamma-trails" are present in juxtaequatorial regions of intrafusal fibers where they exhibit diffuse enzymatic staining activity. "Gamma-plates," in contrast, are located in polar regions of intrafusal fibers and show more discrete enzyme activity in the form of a distinct subneural apparatus at each nerve ending.

In addition to the "gamma-trail" ending, Barker's group [2,10, 17] has also described two forms of the "plate" ending which have

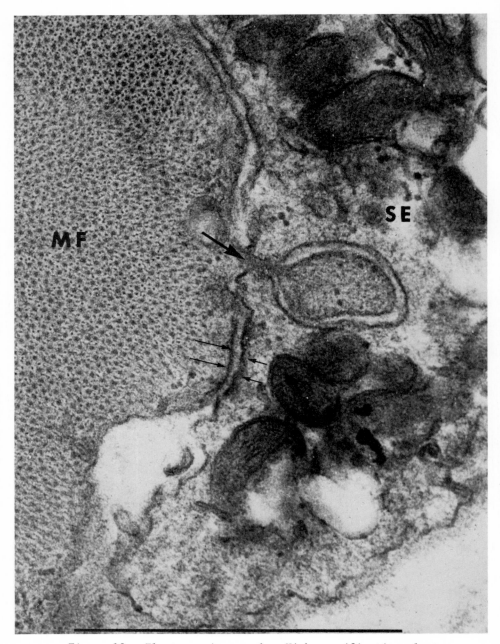

Figure 13. Electron micrograph. High magnification electron
micrograph of a portion of a sensory nerve ending (SE) terminating
on the surface of an intrafusal muscle fiber. The sarcoplasm of
the muscle fiber directly under the nerve ending is completely occupied by contractile myofilaments (MF). Note the partial invagina-
tion of an extension of the muscle cell (thick arrow) into the
center of the nerve ending. A fairly constant intervening cleft is

been termed "p_1-plates" and "p_2-plates." According to them, "p_1-plates" are morphologically identical to extrafusal end-plates and are supplied by alpha motor nerve fibers (i.e., beta-innervation). On the other hand, "p_2-plates" are supplied by gamma motor nerve fibers and degenerate at a much slower rate than "p_1-plates" following nerve section.

Controversy still exists concerning the pattern of distribution of the different forms of motor nerve ending on the two types of intrafusal fiber and their connections to fusimotor axons. Boyd's group [24,27] contends that the nuclear bag and the nuclear chain fiber in the cat receive separate and selective fusimotor innervation. According to them, the nuclear bag fiber is usually innervated by "gamma-plates" while the nuclear chain fiber is innervated exclusively by "gamma-trails." On the other hand, Barker's group [15,17] maintains that motor nerve terminals in the cat are not selective in their distribution on the two intrafusal muscle fibers. According to them, both the nuclear bag and the nuclear chain fiber receive each type of motor ending and receive no selective fusimotor innervation. Work in rat lumbrical muscles, however, indicates that each spindle receives nonoverlapping innervation from two motor axons [142]. Nuclear bag fibers are innervated by one axon, while nuclear chain fibers are innervated by the other axon. Both types of intrafusal fiber, however, were reported to receive two forms of motor ending (see also Mayr [125]).

The controversy over the innervation of mammalian intrafusal maintained between nerve terminal and muscle cell. It is characteristically devoid of basal lamina. In addition, a specialized area of intimate contact (a zonula adhaerens) is present between the limiting membranes of nerve and muscle (thin arrows). Note the increased densities in the two limiting membranes which face this site of contact. An abundance of mitochondria characterizes the sensory nerve terminal. Stained with uranyl acetate and lead citrate. X 85,550.

muscle fibers is almost purely anatomical. The weight of functional
evidence indicates that in fact the two types of fiber, the nuclear
bag and the nuclear chain, are controlled separately by separate
fusimotor axons (the "static" and the "dynamic" axons [32,92,121]).
The anatomical evidence, however, cannot be merely ignored. It is
possible that the explanation lies with the trophic interaction of
nerve and muscle and here two additional studies may be relevant to
the question. Barker and Ip [15] have described a process of
sprouting and degeneration of motor terminals in mammalian muscle.
They contend that motor end plates are being continuously "turned
over." In addition, Mark and his co-workers [118] have shown that
a muscle will not respond to "foreign" innervation in the presence
of "correct" innervation even though an anatomical connection between the foreign nerve and muscle exists. Thus, we suggest that
the controversy could be resolved along the following lines. (a)
End plates on intrafusal muscle fibers are undergoing a continuous
process of sprouting, degeneration, and reformation. (b) Connections,
we assume, are made randomly among the intrafusal muscle fibers,
thus giving rise to the anatomical picture as Barker sees it.
(c) If, however, the process described by Mark et al. [118] can be
extended to muscle spindles, then only the "correct" innervation
will be functionally intact.

    Ultrastructural features of mammalian "plate" and "trail"
endings have been described by several workers in the cat [2,17,42],
in the rat [59,83,109,137], and in man [77]. The two forms of motor
nerve endings differ mainly in the extent of myoneural surface
contact (i.e., in the presence and development of junctional folds)
and in the complexity and appearance of the subneural apparatus
[137]. Figure 14 illustrates a "plate" ending in the polar
region of a nuclear chain fiber. A primary synaptic cleft and
several unbranched junctional folds are present at the neuromuscular junction. The subneural sarcoplasm is well developed
and contains a variety of sarcoplasmic organelles, inclusions, and
a sole-plate nucleus. Numerous clear rounded synaptic vesicles,

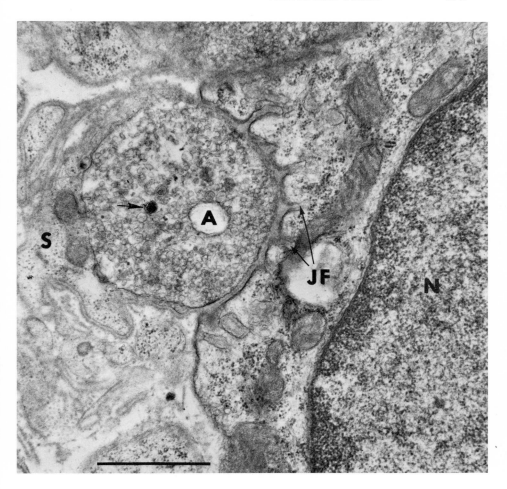

Figure 14. Electron micrograph. Longitudinal section of a "plate-ending" terminating in the polar region of a nuclear chain fiber. Note the presence of regular, unbranched junctional folds (JF) at the myoneural junction. The terminal axon (A) contains an abundance of clear round synaptic vesicles and an occasional dense-core vesicle (arrow). A process of a Schwann cell (S) intimately apposes the outer surface of the terminal axon. Basal lamina is present within the synaptic cleft. The subneural sarcoplasm of the muscle fiber directly under the "plate" ending is well developed

occasional dense-core vesicles, and mitochondria (not shown)
characterize this type of axon terminal. Figure 15 illustrates a
"trail" ending in the juxtaequatorial region of a nuclear bag fiber.
Such an ending reveals the complete absence of junctional folds and
hence a relatively small surface area of neuromuscular contact.
The subneural sarcoplasm is poorly developed; sole-plate nuclei and
an abundance of other organelles and inclusions are lacking. Clear
round synaptic vesicles which predominate in "plate" terminals are
only occasionally encountered in "trail" terminals [137]. Instead,
numerous flattened vesicles and glycogen particles (which are
not found in "plate" terminals) characterize "trail" terminals.

2. *Reptilian Intrafusal Fibers*
a. *Tortoise*

Muscle spindles of the tortoise are simpler than those of the
mammal in that they contain only one type of intrafusal muscle fiber
[48]. Crowe and Ragab [48] were unable to distinguish myofibrillar
differences between the intrafusal and the surrounding extrafusal
fibers of the tortoise. In both of these fibers the T-system and
SR had a similar structural pattern, and triads were situated at
the A-I junctions of each sarcomere. In addition, M bands were
present in all the intrafusal fibers examined. As in the mammal,
mitochondria of tortoise intrafusal fibers are situated between the
myofibrils and run parallel to the longitudinal axis of the muscle
fiber. Close appositions between adjacent intrafusal fibers, like
those described between nuclear chain fibers in the mammal [1,41,
137], are also present in the tortoise. Sensory nerve terminals
ending on tortoise intrafusal fibers are similar in structure to
those described previously in the mammal [109]. In contrast to
mammals, however, no areas of intimate membrane contact between the

---

and is packed with a variety of organelles and inclusions. A sole-
plate nucleus is indicated (N). Stained with uranyl acetate and
lead citrate. X28,500.

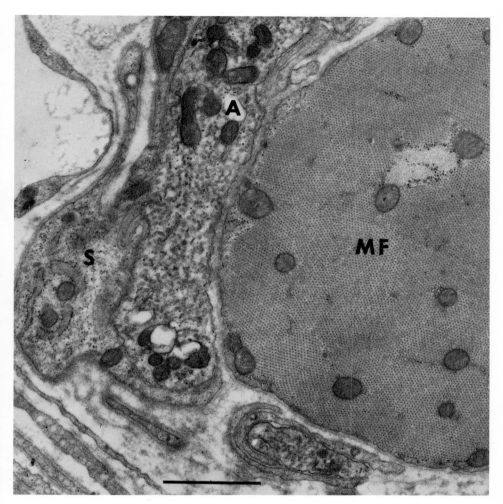

Figure 15.   Electron micrograph.   Transverse section of a
"trail" ending terminating in the juxtaequatorial region of a nuclear bag fiber.   Junctional folds are lacking in the area of myoneural contact, and the terminal axon (A) merely sits on the external
surface of the muscle fiber.   Numerous flattened synaptic vesicles
are present in the terminal axon, while clear round synaptic vesicles
are considerably less abundant.   Mitochondria, numerous glycogen
particles, and a few dense-core vesicles are also present in the
terminal axon.   An attenuated Schwann cell process (S) closely

apposing membranes of muscle fiber and sensory nerve terminal have
been reported in the tortoise.

b. *Snake*

While snake spindles are known to contain only a single intrafusal muscle fiber, Fukami and Hunt [69] have recently described
two distinct types of muscle spindle. The two structural types of
spindle have, in addition, been correlated with "tonic" and "phasic"
behavior [68]. In one type of spindle (i.e., the "phasic" type),
the intrafusal fiber shows a marked enlargement of its equatorial
zone. The other type of spindle (i.e., the "tonic" type) contains
an intrafusal fiber which is more slender and exhibits a thinner
sensory region. The polar regions of both kinds of intrafusal
fiber in the snake exhibit ultrastructural characteristics similar
to mammalian nuclear bag fibers. The myofibrils are tightly packed,
M bands are lacking or inconspicuous, dyads and triads of the
sarcotubular system are sporadic, mitochondria are longitudinally
oriented, and Z bands are thicker than those of the surrounding
extrafusal fibers.

Sensory regions of the "phasic" type of intrafusal fiber contain a paucity of myofilaments and a well-developed central sarcoplasmic core containing a variety of organelles and inclusions.
Similar regions of the "tonic" type of intrafusal fiber, in contrast,
are more slender and contain a relative abundance of myofilaments.
Sensory nerve terminals on reptilian intrafusal fibers possess
ultrastructural features similar to those of the mammal. A 200 Å
intercellular cleft separates the membranes of intrafusal fiber and
sensory nerve terminal with no intervening basal lamina. No evidence
invests the outer surface of the terminal axon. Basal lamina is
encountered in the synaptic cleft. Myofilaments (MF) predominate
in the poorly developed subneural apparatus. Sole-plate nuclei are
lacking and the presence of other organelles and inclusions are less
conspicuous under the "trail" ending. Compare with Fig. 14.
Stained with uranyl acetate and lead citrate. X 25,200.

of intimate membrane contact or specialization between the two
apposing membranes was found.

3. *Amphibian Intrafusal Fibers*

Physiological evidence clearly points to the presence of two
types of intrafusal fiber in muscle spindles of the Anura [64,65,
100,154,155]. These two types appear to be separately innervated
by the large and small motor nerve systems [64,65,75,154,155].
Barker and Cope [13] have reported that there are separate populations of large and small diameter fibers in frog spindles and Smith
[154,155] has related the differences in diameter to the physiological properties and innervation of the fibers in Xenopus laevis.
The larger fibers were innervated by the large motor nerve system
and the smaller fibers by the small motor nerve system [154]. Page
[139] has described two ultrastructural types of fiber in spindles
from the frog Rana temporaria. One type had a striation pattern
resembling a "fast" extrafusal fiber, while the other had an appearance reminiscent of the nuclear bag fiber in the mammal, that is,
the M band was either absent or poorly defined. Page [139] called
this second type of fiber an "intermediate" fiber since in some
respects it was structurally intermediate between the extrafusal
fast and slow varieties of fiber [138]. Both types of intrafusal
fiber differed from extrafusal fibers in some ways. In the "fast"
variety the T-tubules tended to lie parallel to the fiber axis and
the sarcoplasmic reticulum was less regularly organized than in
extrafusal "fast" fibers. The "intermediate" fiber, although it
lacked an M band (as does the extrafusal slow fiber in the frog),
had I filaments entering the Z band in an orthogonal array (unlike
the "slow" fiber). The sarcoplasmic reticulum in these fibers was
not seen to form the "fast" type of collar around the myofibril at
the H zone. Page [139] found that the end plates on the "fast"
type of intrafusal fiber resembled those on "fast" extrafusal muscle
fibers [96,138], while the motor terminals on the intermediate
fibers were of the "en grappe" type characteristic of small-nerve
endings [138]. The occurrence of the "fast" and "intermediate"

types of fiber has been confirmed in spindles from Xenopus (Smith and Page; Smith, Malhotra and Nag, unpublished observations). Thus, although Page [139] herself did not observe any difference in fiber diameter between the two types, it is reasonable to suppose that the functionally described "small" fiber [154,155] corresponds to the "intermediate" fiber described by Page while the "large" fiber [154,155] corresponds with the fiber having a "fast" structure.

Barker [11] has suggested that muscles which receive no small motor innervation to the extrafusal muscle (e.g., the sartorius muscle) may contain spindles that lack those intrafusal fibers that usually are innervated by the small motor system. He made this suggestion because it is well known that intrafusal muscle fibers in the frog are innervated by branches of axons supplying extrafusal motor units [64,75,100]. No published structural work relevant to this question exists, although some unpublished work by Page and Smith did fail to reveal any "intermediate" fibers in sartorius muscles from Xenopus. Functional studies by Brown [31] do indicate that there is a difference in the types of intrafusal muscle fibers in spindles from muscles which receive a small motor nerve supply and those which do not.

Thus far we have considered the evidence that two structural and functional types of muscle fiber occur in anuran muscle spindles. The point has been dwelt upon because most of the ultrastructural descriptions of intrafusal fibers in the frog do not take this reasonably well-established fact into account. Consequently, in what follows we are unable to relate the ultrastructural features to any particular fiber type.

The first ultrastructural studies of intrafusal fibers in the Anura revealed the presence of structural specializations in the area of sensory innervation [102,144]. In this region the muscle fibers became fenestrated and lost much of their contractile material; this region, which is about 50-100 μm long, was named the "reticular zone" [102] to distinguish it from the "dense" polar zones. Subsequent work has shown that the reticular zone has a periodic structure [95]. The reticular zone exhibits alternate widening and thinning with a periodicity of about 1.5 times the sarcomere length. Karlsson [95] has recently given the name "reticulomere" to the unit periodic structure in the reticular zone. Single isolated sarcomeres occur in each of the widenings of the reticulomere, attaching by what appears to be Z band material to the sarcolemma at each end [98]. These points are illustrated in Fig. 16. Whether each type of fiber, the large and the small, or the fast and the intermediate, contains similar reticular zones is not known. There is' some evidence [102,157] that the small variety of intrafusal fiber may contain no such modified region.

Leptomeres, similar to those described in mammalian intrafusal fibers, occur in intrafusal fibers of the frog in the region of sensory contacts [97,102]. Two types of leptomeres have been described: those with a periodicity of about 1870 Å (Type I) and those with a periodicity of 300 Å (Type II) [97]. The type I leptomeres show changes in length which follow the sarcomere length of the fiber [97,102]. Whether the leptomeres are in series or in parallel with the sarcomeres is, however, not certain.

As is the case for intrafusal muscle fibers in other vertebrates the sensory region is filled with a variety of organelles: nuclei, mitochondria, Golgi membranes, rough-surfaced endoplasmic reticulum, granules, vesicles, tubules, and filaments. Some of these organelles are indicated in the diagram of Fig. 16. Andersson-Cedergren and Karlsson [4] believe that a number of the small particles are ribosomes; these may be strung together in rows as polyribosomes. This finding may indicate that the sensory region of the intrafusal fiber is the site of active protein synthesis.

No special modification of the frog intrafusal fiber has been shown to occur at its junction with the sensory nerve endings. Katz [102] thought that some structure might span the 150 Å gap between the sensory endings and the muscle fiber. Karlsson et al.

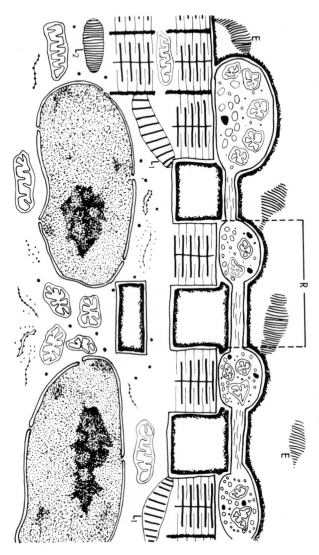

Figure 16. Diagrammatic representation of the "reticular"
(sensory) zone of an amphibian intrafusal muscle fiber. A series
of centrally placed nuclei and a central core of sarcoplasm rich in
a variety of organelles characterizes the interior of this region
of the muscle fiber. The periphery of the muscle fiber contains a
series of periodic surface projections termed "reticulomeres" (R).

[98] merely state that the apposition is "intimate." In this laboratory we have obtained no evidence that any structure crosses the
gap. As with the mammalian spindle, the basal lamina does not
enter the interface between sensory ending and muscle fiber.

Recent work in this laboratory (Smith, Malhotra, and Nag, unpublished) has shown that periodic extracellular structures, resembling at first glance the Type I leptomere, are arrayed between the
sensory endings and between sensory endings and intrafusal muscle
in Xenopus (Fig. 16). Karlsson [95] has recently reported periodic
extracellular structures in frog spindles, i.e., "exomeres."
However, these structures are seen only in maximally stretched
spindles whereas the periodic extracellular structures in Xenopus
may be seen in relaxed specimens. The function of this structure,
as is the case for many of the intracellular organelles, is completely
unknown.

III.   HISTOCHEMISTRY

In recent years it has been possible to classify the extrafusal
muscle fibers of most vertebrates into at least three categories by
using histochemical stains. The high myoglobin content of the so-
called "red" muscle fiber of the mammal gives it its characteristic
color. In addition, this kind of fiber exhibits a high oxidative
(mitochondrial) enzyme activity, a low glycolytic (anaerobic) enzyme
activity, and a low myofibrillar ATPase activity [58,160]. The
"red" fiber characteristically is of small diameter. Conversely,
Each reticulomere contains a single isolated sarcomere attached on
both sides to the sarcolemma. Interconnected sensory nerve bulbs
can be seen terminating on each reticulomere beneath the basal
lamina of the muscle fiber. Two types of leptomeric organelles,
termed Type I (L_1) and Type II (L_2), are present in the sarcoplasm
of the muscle fiber. In addition, several periodic extracellular
structures, termed "exomeres" (E), are seen in close association to
the sensory nerve endings and to the sarcolemma of the muscle fiber.

the "white" muscle fiber, which is larger in diameter, has a low
myoglobin content, low oxidative enzyme activity, high glycolytic
enzyme activity, and a high myofibrillar ATPase activity. Between
these extremes one can distinguish muscle fibers with intermediate
levels of enzyme activities; these have been termed "intermediate"
or "pink" fibers [70]. The histochemical profiles of mammalian
muscle fibers may not be applicable to other vertebrates; the precise pattern described above certainly does not extend to amphibian
muscle [62,111].

Just as it has been possible to demonstrate different kinds of
extrafusal muscle fibers by histochemical techniques, so can one
distinguish different kinds of intrafusal muscle fibers. For
example, it has been shown in a variety of mammals that the small
diameter nuclear chain fiber has a greater oxidative enzyme (SDHase)
activity than the larger nuclear bag fiber: monkey [131], mouse,
rat, cat, and human [133], rat [172], human [159]. A greater
mitochondrial ATPase activity (cat [82]) and a more intense Sudan
Black B staining (rat [125]) occurs in nuclear chain fibers than in
nuclear bag fibers. In addition, the nuclear chain fiber exhibits
a greater myofibrillar ATPase activity than does the nuclear bag
fiber (Fig. 17): monkey [131], cat [132], human [159]. James [91]
has found that in a number of mammals the nuclear chain fibers are
myoglobin-poor while the nuclear bag fibers are myoglobin-rich. It
is apparent from the foregoing that the "histochemical profile" of
individual intrafusal muscle fibers in the mammal does not match
the profiles of extrafusal muscle fibers. In addition, it is worth
noting that the levels of oxidative enzyme activity in intrafusal
muscle fibers may be much higher than in any extrafusal fiber.
Nyström [132] has shown that $NADH_2$-tetrazolium reductase is so
prevalent as to cause the fibers to appear completely black (Fig.
18).

A third type of mammalian intrafusal muscle fiber, intermediate
in histochemical staining intensity and diameter to the two general
types described above, has been reported in the rat, guinea pig,

rabbit, and human (SDHase) [133], in the dog (myoglobin) [170], and
in the rabbit (SDHase, myofibrillar ATPase, and phosphorylase) [12,
16].

In the toad, Nyström [132] has identified two types of intrafusal muscle fiber on the basis of their staining reactions for
$NADH_2$-tetrazolium reductase.   The small intrafusal muscle fibers
stained darkly while the large fibers stained sparsely.   Both types
of fibers stain darkly for myofibrillar ATPase [Fig. 19(a)].   While,
as implied above, it is rash to equate the histochemistry of intrafusal fibers with that of extrafusal fibers, it may be worth noting
here that neither of the intrafusal fibers in the toad has the
characteristic histochemical profile of the slow extrafusal muscle

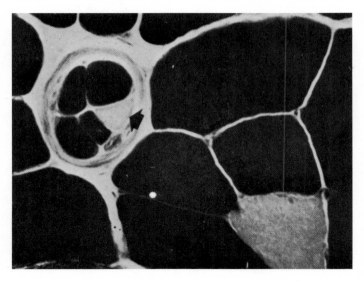

Figure 17.   Transverse section of a portion of rabbit skeletal
muscle stained for myofibrillar ATPase.   Two types of intrafusal
fibers can be seen in the polar region of a muscle spindle.   One
light-staining fiber (arrow) and four dark-staining fibers are
enclosed within the spindle capsule.   In this case [158] the fibers
were not identified as nuclear chain or nuclear bag.   In other
mammals, nuclear chain fibers stain more darkly for myofibrillar-
ATPase.   In addition, two types of extrafusal fibers are present.
X 520.   (Modified from Spiro and Beilin [158].)

fiber [62,111] (Fig. 19). In the chick, Germino and D'Albora [71]
found that all intrafusal fibers were rich in oxidative enzymes.
Intrafusal muscle fibers in the adult chicken have, however, been
shown to be separable into at least two types by a variety of
staining methods [143].

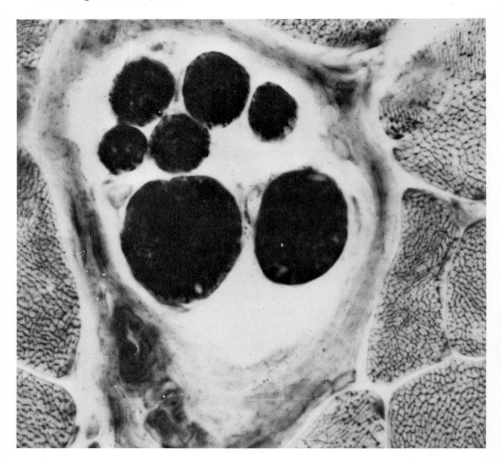

Figure 18. Comparison of oxidative enzyme levels in intra- and
extrafusal muscle fibers in a hind limb muscle from the cat. The
intrafusal muscle fibers stain very darkly and uniformly for $NADH_2$-
tetrazolium reductase, whereas the extrafusal fibers stain only in
the intermyofibrillar spaces. (Section and photograph by B. Nyström.)

When attempts are made to compare the results of histochemical studies of extrafusal fibers with those obtained for intrafusal muscle fibers, it must be kept in mind that, unlike extrafusal fibers which are morphologically and histochemically uniform along their lengths [66], intrafusal fibers are known to show marked structural alterations along their lengths [9,128]. It is possible that a histochemical profile of an intrafusal muscle fiber in one part of a spindle, such as the polar zone, may be quite different from the profile of the same fiber in another part of the spindle such as the equatorial zone. Indeed, Yellin [172] has found that in rat spindles, the difference in the enzyme patterns shown by nuclear bag and nuclear chain muscle fibers diminishes as sections are taken closer to the equatorial zone.

Recently, Guth and Samaha [78] have reported that mammalian skeletal muscle contains two forms of myofibrillar ATPase. Any given extrafusal muscle fiber contains only one form of the enzyme. The alkali-stable, acid-labile form was found in the "white" "fast-twitch" type of muscle fiber while the acid-stable, alkali-labile form was found in "red" "slow-twitch" muscle fibers. In contrast Yellin [173] has shown that nuclear bag and nuclear chain muscle fibers in the rat may each possess both forms of myofibrillar ATPase. He observed that this "dual ATPase activity" was uniform along the intrafusal fibers from the juxta-equatorial to the extreme polar regions of the spindle, but all the fibers were generally nonreactive in the equatorial region. Since it has been shown that myofibrillar ATPase activity in muscle is neurally regulated [79,99,147,148], Yellin [173] attributed the dual ATPase activity of intrafusal fibers to multi-axonal efferent innervation. However, while they may possess multiple end plates, it has not been shown that intra-fusal muscle fibers in the rat do receive a multi-axonal innervation. Evidence exists [142] that intrafusal fibers in the rat's lumbrical muscle are innervated in a very simple, uni-axonal, fashion. The possibility also remains that the histochemical characteristics of intrafusal muscle fibers are partly under the control of their sensory innervation.

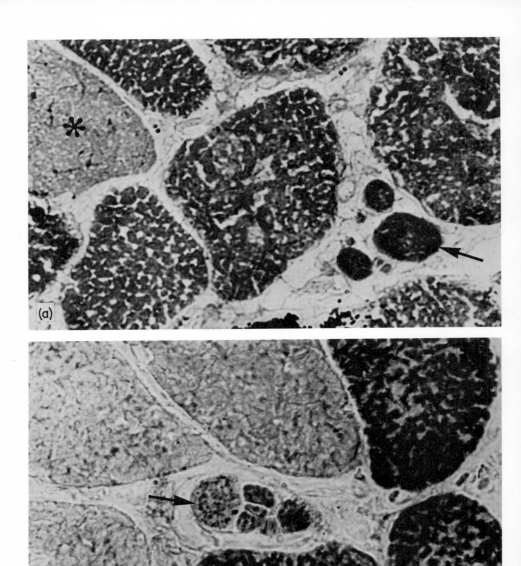

Figure 19. (a) Transverse section of a portion of the tibialis
anterior muscle of the toad <u>Xenopus</u> <u>laevis</u> stained for the demonstration of myofibrillar ATPase activity. Three intrafusal muscle
fibers, two small and one large (arrow), stain uniformly and more
intensely than any of the surrounding extrafusal muscle fibers.
One pale slow fiber (asterisk) and several dark staining extrafusal

An unusual finding, which is not easily related to any other histochemical observation, but which nevertheless emphasizes the distinct reactions of intrafusal fibers, is that these fibers in the frog have a very high affinity for Procion yellow [141]. We are not aware that a similar affinity for this fluorescent dye has been demonstrated for mammalian intrafusal fibers.

IV. DEVELOPMENT

During the histogenesis of vertebrate striated muscle three successive stages of muscle development have been described [104, 161,165,178] (Fig. 20). The first is the "myoblast stage" during which nondifferentiated mesodermal cells form elongated mononucleated myoblasts [Fig. 20(A)]. The second stage is the "myotube stage," characterized by the fusion of several myoblasts, the accumulation of nuclei in the center, and the formation of myofibrils at the periphery of each new myotube [Fig. 20(B)]. The third stage of development may pursue either of two separate courses [Fig. 20(C)]. The majority of myotubes differentiate into prospective extrafusal muscle fibers by the peripheral migration of nuclei and the further development and proliferation of myofibrils within each new cell. A small number of myotubes, on the other hand, contain nuclei which do not migrate to the periphery of each cell but, instead, bunch together in areas which will eventually form equatorial regions of future intrafusal muscle fibers.

---

fibers are noted in the field. X 800. (b) Transverse section of a portion of the iliofibularis muscle of Xenopus laevis stained for the demonstration of succinic dehydrogenase activity. A muscle spindle, containing several small diameter and one large diameter (arrow) intrafusal fiber, is noted in the center of the field. The small intrafusal fibers stain more intensely than the large intrafusal fiber. None of the intrafusal fibers, however, resembles the three clear (slow) extrafusal fibers on the left. X 800.

Certain aspects of the development of intrafusal muscle fibers and their innervation have been reported in birds [52,130,149,165], in lizards [116], and in a variety of mammals [28,44,49,50,54,86, 94,110,113,117,163,171,175,178]. No study exists of the development of amphibian intrafusal fibers.

Intrafusal muscle fibers begin to develop in embryonic life, i.e., day 11 in chick embryos [165], the fourth month in human embryos [50], seven days before birth in the rabbit [177], and three days before birth in the rat [175]. At birth the presence of two well-differentiated intrafusal fiber types differing in size and nuclear arrangement [177] and phosphorylase activity [171] has been reported in rodents.

It has been shown that a progressive increase in the number of intrafusal fibers occurs during the early postnatal period in the rat [117,175]. At birth two intrafusal fibers are distinguishable, a nuclear bag and a nuclear chain, and by day 10 the adult number of fibers (i.e., four to five in the rat) is present [117]. The process by which intrafusal fibers increase in number has been disputed for several years and several mechanisms have been suggested [44,50,113,117]. Latyshev [113] maintained that newly formed intrafusal fibers develop de novo from elements of the inner layer of the spindle capsule. He presented little evidence, however, to

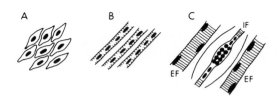

Figure 20.   Stages in the development of muscle (after Zelena and Hnik [178]). A, myoblast stage. B, myotube stage. C, adult stage showing extrafusal fibers, EF, with peripheral nuclei, and an intrafusal fiber, IF, with an equatorial accumulation of nuclei.

support this view. Couteaux [44] believed that new intrafusal fibers developed by multiplication and fusion of myoblasts which were situated in close proximity to the surface of existing myotubes. Marchand and Eldred [117] studied the postnatal development of intrafusal fibers in the rat; they rejected Couteaux's suggestion after finding no evidence of increased mitotic activity among the "periaxial sheath cells" of the spindle during the early postnatal period (i.e., when intrafusal fibers are known to increase in number). Whether the "parent" intrafusal fibers, i.e., the nuclear bag and the nuclear chain, develop prenatally by the multiplication and fusion of myoblasts has not yet been critically examined.

It has been suggested that intrafusal fibers might increase in number by the longitudinal division (splitting) of pre-existing muscle fibers [50]. Splitting, bifurcation, branching, or fusing of intrafusal fibers in adult muscle spindles has been reported [7, 14,18,22,24,67,81,169]. In examining serial sections of spindles of the neonatal rat (i.e., days 0 to 10), Marchand and Eldred [117] frequently found longitudinal splitting of individual intrafusal fibers in addition to a staged increase in the number of nuclear bag and nuclear chain fibers with age. They interpreted such a longitudinal division of fibers as the principal mechanism for their postnatal increase. In addition, they suggested that the branching of adult intrafusal fibers previously reported by other workers (see references above) may "represent a failure of the splitting process to go to completion."

The mechanism by which the adult complement of nuclei is attained in developing intrafusal fibers has recently been investigated in neonatal rats [28,117]. Earlier studies on the prenatal development of muscle spindles indicate that intrafusal fiber nuclei "divide intensively" in rat embryos [175] or undergo "rich amitotic division" in chick embryos [165]. Marchand et al. [117] (using conventional histology and ^{3}H-thymidine incorporation techniques) and Bravo-Rey et al. [28] (using ionizing radiation and colchicine administration) have presented evidence that the total complement

of nuclei is present in the two "parent" intrafusal fibers at birth.
At the time of longitudinal splitting of "parent" intrafusal fibers,
the original nuclei, they suggest, are merely distributed to the
"daughter" fibers without further duplication.

The role of sensory innervation in the development of the
muscle spindle has been investigated in the rat [175,178]. It is
well known that future intrafusal fibers develop in close proximity
to sensory nerve endings. During the early stages of differentiation (i.e., the myotube stage), primary (Ia) sensory endings are
the only nervous elements present in the developing spindle. Secondary sensory (IIa) and fusimotor nerve fibers do not reach the
spindle until after birth [175]. Sensory innervation has been
shown to induce the formation and further differentiation of intrafusal fibers during the initial stages of spindle development.
Experimental denervation of muscles at the myotube stage in rat
embryos, for example, prevents the formation of intrafusal fibers
in the denervated muscle [175]. Denervation at birth, on the other
hand, prevents the further development of pre-existing spindles and
leads to a reduction in their number, retarded growth, and atrophy.

The influence of fusimotor innervation on the differentiation
of intrafusal fibers has been studied in neonatal rats [177]. Although muscle spindles continue to differentiate and are not reduced
in number after de-efferentation (ventral root section), the intrafusal fibers are reduced in size and undergo atrophy.

V. PATHOLOGY AND EXPERIMENTALLY INDUCED CHANGES

Since the initial experimental studies of Sherrington [153] and
Batten [18] it had generally been thought that, in contrast to
extrafusal muscle fibers, intrafusal fibers remain relatively
resistant to structural alterations for long periods of time after
nerve section; see also [3,162]. Tower [166], on the other hand,
demonstrated marked atrophy and degeneration of intrafusal fibers
after selective denervation of adult cat hind limb muscles. She

showed that removal of dorsal root ganglia resulted in degeneration of intrafusal fiber equatorial regions while ventral root sectioning resulted in atrophy of intrafusal fiber polar regions. She concluded, therefore, that both motor and sensory nerve fibers maintain the structural integrity of the intrafusal muscle fibers. Boyd [23, 24] confirmed this view and in addition showed that sectioning the ventral roots caused nuclear chain fibers to atrophy more rapidly and to a greater extent than nuclear bag fibers. Atrophy of intrafusal fibers, however, is generally less extensive and occurs considerably later than that of extrafusal fibers after nerve section [3,23,24,80,162].

It was initially thought that complete denervation of a muscle eventually led to disintegration of the intrafusal fibers with a substantial reduction in the number of spindles [166]. It has recently been shown in the rat, however, that although the intrafusal fibers undergo atrophic changes after total denervation, the average number of spindles is not reduced up to two years after the operation [80]. In addition, it has been shown that tenotomy results in a marked atrophy of extrafusal fibers, while intrafusal fibers do not atrophy although they are somewhat reduced in length [174,177].

Relatively little is known about specific pathological changes in intrafusal muscle fibers since it had long been suspected that muscle spindles persisted, apparently unaffected, in many muscular diseases [38]. Difficulties in evaluating discrete pathological changes in muscle spindles from random human biopsy specimens are obvious. Recently, however, some attention has been directed to the involvement of muscle spindles in various neuromuscular diseases in rodents and humans [34,51,112,127,140,152].

Patel and co-workers [140] found no specific pathological changes in intrafusal fibers in human muscular dystrophy, dystrophia myotonia, or chronic polymyositis. Daniel and co-workers [51], in contrast, reported a substantial increase in the number of intrafusal

fibers (up to 60 per spindle!) in human dystrophia myotonia. In an
extensive study of muscle spindle pathology Cazzato and Walton [34]
concluded that the degree and extent of structural alteration in
intrafusal fibers in a variety of human neuromuscular disorders
(i.e., denervation atrophies, progressive and congenital muscular
dystrophy, polymyositis, etc.) were related to the time of onset
and stage of the disease. In addition, Meier [127] maintained that
intrafusal fibers are generally "disease resistant" or may undergo
minor changes, if at all, during the late stages of muscle disease.

VI. MECHANICAL PROPERTIES

A. Passive Mechanical Properties

The sensory discharge from a muscle spindle may reflect the
steady-state length of the muscle or it may reflect the rate of
change of length of the muscle. The primary ending of the mammalian
muscle spindle, for instance, can respond both to the steady length
of the muscle and to its rate of change. The earliest explanation
of the origin of the rate sensitivity of muscle spindles was that
of Matthews [119,120] who proposed that if the intrafusal muscle
fibers were less viscous in the region of their sensory endings
than at their poles, then the sensory discharge would reflect the
rate sensitive mechanical properties of the intrafusal muscle fibers.
We refer to this class of ideas as the "mechanical hypothesis."
With the advent of the length servo hypothesis [129] for the role
of the muscle spindle in the control of movement, and with the
demonstration of the independent modulation of the rate-sensitive
and steady-state discharges of the mammalian primary ending through
"dynamic" and "static" fusimotor stimulation [92,121], the mechanisms responsible for rate sensitivity became of great interest.
Figure 21, after PBC Matthews [122], illustrates the basic idea.
The nucleated (equatorial) region of the nuclear bag muscle fiber
could be regarded as a purely elastic element since it lacked
contractile material; the poles of the fiber are supposed to have

viscous properties, attributable to the contractile elements, as
well as elastic properties [Fig. 21(A),(B)]. An elongation of the
spindle would cause a velocity-sensitive elongation of the elastic
(nuclear bag) element. The length of the purely elastic element
would subsequently decrease, with a time constant which would depend
on the values of the viscous and elastic elements, to some length
such that the forces in the extended elastic elements balance. The
frequency of the sensory discharge is supposed to follow the length
changes in the nuclear bag region [Fig. 21(D)].

The mechanical hypothesis has been expressed quantitatively by
a number of workers [6,45,46,72,89,145,167]. The simple first-order
model given above has been extended to a second-order model by
Rudjord [145]. It is now unfortunately true that the theoretical
treatments have far outrun experimental evidence. Further elabora-
tion of such models can serve no purpose until some means is found

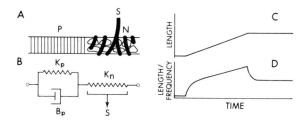

Figure 21. Illustration to show the basis for the "mechanical
hypothesis" on the origin of rate sensitivity in the muscle spindle.
In A, a portion of an intrafusal fiber is diagrammed. P, polar zone;
N, nucleated region; S, sensory ending. Below, in B, the mechanical
properties of these regions are shown as a lumped-parameter model.
K_p and B_p, stiffness and viscosity of the polar region. K_n, stiffness of the nucleated region. The sensory endings, S, are supposed
to respond to length changes in the nucleated region. If the muscle
were to be stretched as in C, then, on this model, the length of the
nucleated region and the frequency of sensory impulses would have a
time course as in D. (After Matthews [122].)

to test them. A few direct observations indicate that, for the
mammalian muscle spindle, the mechanical hypothesis on the origin
of rate sensitivity could be essentially correct. Smith [156] has
presented measurements which show that in rat spindles which were
rapidly stretched, the behavior of the nuclear bag region was qualitatively what would be expected from the hypothesis. The nuclear
chain fiber, on the other hand, showed no region which shortened
following rapid elongation. This too is expected from the low rate
sensitivity of secondary endings [122]. Boyd [25,26] has described
similar mechanical properties for the nuclear bag fibers of cat
tenuissimus spindles; the slow shortening of the nuclear bag region
following a rapid extension of the spindle is shown very well in
his motion pictures. It remains true, however, that the experimental
observations, while suggestive, are so few and so qualitative that
judgment will have to be reserved on the applicability of the mechanical hypothesis to the mammalian spindle.

Theory and experiment are better balanced in studies of the
frog muscle spindle, although the outcome has been surprising.
Experimenters working with the frog spindle have been fortunate in
that a simple, yet realistic, mechanical model of the intrafusal
muscle fiber has been developed by Houk, Cornew, and Stark [89].

As in other muscle spindles, the flow of events connecting the
change in length of the muscle, or contraction of the intrafusal
muscle fibers, with the final output in the sensory nerve is assumed
to be represented by the diagram of Fig. 22.

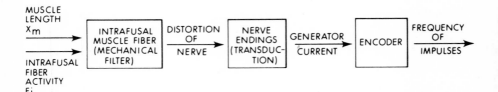

Figure 22. Schematic flow-diagram of the transduction process
in a muscle spindle (after Houk et al. [89]).

This diagram has considerable conceptual importance since the properties of the intrafusal muscle fibers have been deduced by controlling one of the inputs to the muscle spindle (X_m or F_i) and measuring either the generator current originating in the sensory endings or the afferent discharge in the sensory nerve. It has been shown, originally by Katz [101] and recently by Ottoson and Shepherd [135], that the frequency of afferent nerve impulses in the frog spindle is instantaneously proportional to the magnitude of the generator current. For the want of evidence suggesting otherwise, it has been assumed that the mechanical to electrical transducing mechanism is also linear so that the magnitude of the generator current is directly proportional to the deformation of the sensory endings. Further, it is implicit in this model that the sensory endings are deformed exactly as is the intrafusal muscle fiber on which they are located. Thus, according to this concept, the sole explanation for an observed rate-sensitive or adaptive behavior in the output of the spindle lies with the mechanical characteristics of the intrafusal muscle fibers. Similar reasoning has been used to explain the adaptive characteristics of mammalian spindles [115]. Houk et al. [89] argue that the total length of the spindle in the frog, that is the length of the intrafusal muscle fibers (X_s) is proportional to the length of the muscle (X_m).
That is

$$X_s = K_m X_m \tag{1}$$

where K_m is a constant less than unity. The total length of the intrafusal muscle fibers may be divided into two parts: the length of the reticular zone (X_r) and the remainder (X_d) thus

$$K_m X_m = X_r + X_d \tag{2}$$

The mechanical properties of the muscle were modeled by a contractile component in parallel with a viscous element B_d (to represent the contractile length of muscle fiber, X_d) and a purely elastic element having a length X_r and a spring constant K_r to represent the reticular zone. These points are illustrated in Fig. 23(A). From this model, the force produced by the contractile part of the intrafusal fiber is

$$F = F_i + B_d\, X_d \tag{3}$$

where F_i is the force generated by the contractile component. This can be equated to

$$F = K_r\, X_r \tag{4}$$

thus

$$F_i + B_d\, X_d = K_r\, X_r \tag{5}$$

Equations (2) and (5) can be used to solve for the changes in length of the reticular zone, ΔX_r, and of the poles of the fiber, ΔX_d, following changes in the length of the spindle, X_s, or changes in the force generated by the intrafusal fibers, F_i. It should be noted that in the equations above the only values which have been measured are the lengths of the reticular and polar parts of the spindle at rest. The remainder of the values must be taken as reasonable estimates. Houk et al. show solutions representing the changes in length of the reticular zone and polar zones of the fiber following a step change in muscle length (X_s). The time course of the changes are shown in Fig. 23(B), here $\Delta F_i = 0$. Since the ensuing steps in the formation of the sensory output of the spindle are assumed to be linear, the use of suitable proportionality constants, together with estimates of the numbers of sensory endings affected by changes in length of the reticular zone and dense zones, should enable one to predict the sensory output of the spindle if the input is known. The line of reasoning is ingenious but tenuous and it is perhaps not surprising that, with suitable selection of parameters, the theoretical predictions follow experimental data quite closely. Examples of such comparisons using a very similar model have been shown by Gottlieb, Agarwal, and Stark [72].

Direct tests of the mechanical hypothesis for frog intrafusal muscle fibers are technically difficult to perform but have been carried out. Ottoson and Shepherd [134] used the isolated capsular

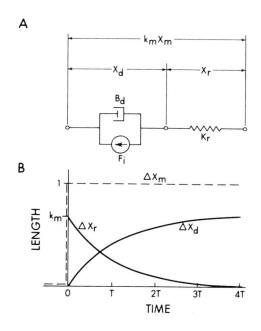

Figure 23. A schematic representation of the mechanics of an amphibian muscle spindle is shown in A. The spindle has a length $K_m X_m$, made up of X_d (the length of the dense, or polar zones) and X_r (the length of the reticular zone). The dense zone has viscous properties, B_d, and can generate force, F_i. The reticular zone is elastic, K_r. A sudden increase in the length of the muscle, ΔX_m, gives rise to changes in the lengths of the dense and reticular zones which are shown in part B. The reticular zone increases in length suddenly and returns, ΔX_r, slowly to its original length. At the same time, the dense zone increases in length, ΔX_d. The changes in length follow an exponential time course. Time, on the abscissa, is plotted as time constants of the exponential process. (Illustration redrawn from Houk et al. [89].)

region of muscle spindles, stretched the specimen rapidly, and
compared the length changes of segments of intrafusal muscle fiber
in the equational region with the wave form of the generator current
produced by similar specimens under similar circumstances. Their
findings were that the changes in length of the intrafusal muscle
fiber in the sensory region followed the waveform of the displacement of the ends of the specimen; no difference in mechanical properties between various regions of the intrafusal fibers was noted.
This experiment does demonstrate that some rate-sensitive behavior
may be attributed to the transduction process. It does not, however,
eliminate the possibility that the intact intrafusal muscle fibers
may have mechanical properties as prescribed by the mechanical
hypothesis. Houk et al. [89] have pointed out that if the majority
of the polar regions of the intrafusal muscle fibers were to be
removed, then the "overshoot" predicted in the lengthening of the
reticular zone would be reduced and its time constant of decay to
the steady-state would be lengthened. These effects could have
prevented Ottoson and Shepherd [134] from observing any rate-
sensitive changes in the length of the sensory regions of the muscle
spindles. The prediction by Houk and co-workers has been used by
Blinston (unpublished work) in this laboratory in order to test
their model indirectly. In fact, if the spindle is held at a length
at which rate sensitivity occurs (see below) then removal of the
polar regions of the intrafusal muscle fibers does not affect the
time constant of decay of the sensory discharge.

    Koles and Smith have also examined the mechanical responses of
intrafusal muscle fibers by direct observation. (The results were
reported in a general fashion at the Muscular Dystrophy Association
Symposium in Cleveland 1970 [157].) In this work intact muscle
spindles were observed with a Nomarski optical system. The reticular zone could be recognized using the following criteria: (a) the
presence of sensory endings at the region, (b) a coarsely banded
appearance, the bands having a period at normal muscle lengths of
about 1.5 times the sarcomere length, (c) the region had a length

approximating the length of the reticular zone, from 50-100 μm [102]
and (d) the same region extended during contraction of the polar
parts of the intrafusal muscle fiber [155]. A study of the steady-
state characteristics of the spindle revealed that the reticular
zone did not extend until the muscle was stretched to about 110% of
its <u>maximal</u> length in the animal. When the reticular zone did
begin to extend, its stiffness was initially much less than that of
the neighboring dense zones (Fig. 24). Evidence was obtained that
extensions of the muscle up to the point at which the reticular zone
began to lengthen caused lengthening only of the polar, extracapsular
regions of the intrafusal muscle. This latter finding was taken to
indicate that the capsule represents a very stiff element in parallel

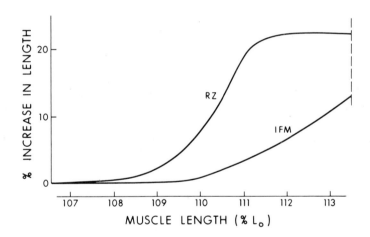

Figure 24. Comparison of the compliances of the reticular
zone and the intracapsular portion of the dense zone of intrafusal
muscle fibers in <u>Xenopus</u> <u>laevis</u>. Plotted on the ordinate are the
percentage increases in length of the reticular zone (RZ) and the
dense intrafusal muscle fiber (IFM) as the muscle is stretched above
its maximal length in the animal (L_o). The dashed vertical line
indicates fracture of the intrafusal muscle fibers at some extracapsular site. (Illustration taken from Ref. [157].)

with the sensory region of the intrafusal muscle fibers. Further
experiments showed that during ramp extensions of the muscle which
caused lengthening of the reticular zone, the sensory discharge
showed a marked rate sensitivity while the length changes in the
reticular zone were not rate sensitive. One must assume then that
viscous forces are not important in describing the mechanical properties of intrafusal muscle fibers in <u>Xenopus</u>. At this level of
experimentation, the spindle may be modelled simply as a system of
elastic elements, as shown in Fig. 25.

The overall conclusion from these attempts to investigate the
mechanical properties of intrafusal muscle fibers in the amphibia
is that the rate sensitivity of the afferent discharge does not
reflect the gross mechanical properties of the fibers. Rate sensitivity will presumably be found to be caused by mechanical structures
not visible with the light microscope, or, it may be a property of
the transduction process [88]. The control of rate sensitivity
would then not be a direct result of the modification of the mechanical properties of the intrafusal muscle. It would rather result
from some process in which the rate sensitivity of the transducing
mechanism is a function of the initial distortion of the sensory
endings. This latter mechanism certainly appears to be most likely
in <u>Xenopus</u> [157].

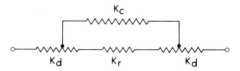

Figure 25. Schematic representation of the mechanical properties of the muscle spindle in <u>Xenopus laevis</u>. The passive spindle
may be modelled as a system of springs. The spring constants are
indicated: K_d, spring constant of dense, or polar, zones of the
intrafusal fiber; K_r, spring constant of reticular zone, and K_c,
the capsule. The springs are nonlinear, but in general $K_d > K_r$,
and the parallel combination of K_c and $K_r > K_d$.

The experiments outlined above are not completely free from controversy. Karlsson [95] has recently described electron microscopic measurements of the lengthening of parts of the reticular zone at various muscle lengths. While he reports, in agreement with the findings of Koles and Smith [157], that the reticular zone is less stiff than the dense zones of the intrafusal muscle fiber, he maintains that the reticular zone will extend at normal muscle lengths. This observation might be explained merely by modifying the elastic properties of the capsule, K_c of Fig. 25. Such a simple explanation, while possible, is not attractive since the implication for the role of the intrafusal fibers in the control of the sensory discharge is quite different in the two cases. The work of Koles and Smith [157] suggests that the spindle only functions as a stretch receptor during contraction of the intrafusal muscle fibers; Karlsson's work [95] suggests that, in a different species of frog, the spindle may function as a stretch receptor while the intrafusal fibers are passive.

Other difficulties in interpretation arise in the fact that the rate sensitivity of the endings is measured in an axon fed by branches which innervate two different kinds of intrafusal muscle fiber. Very little account has been taken of this in the work discussed so far; it should be noted that the work of Matthews and Westbury [124] implies that rate sensitivity is under the control of the small motor nerve system and not of the large motor nerve system. Certainly, the role played by the intrafusal muscle fibers in the control of the sensory discharge is by no means clear.

A recently described property of extrafusal muscle in the frog, the "short range elastic component" [87], may have great significance in the function of intrafusal muscle. Hill [87] described a rapid, early increase in tension during passive extension of the sartorius muscle. The rapid development of tension occurred over extensions of up to 0.2% of the muscle length, and was attributed to the resistance offered by stable bonds between actin and myosin filaments. The tension developed in this way is not large, and a simple

calculation will show that it is not likely to be a directly measurable effect in intrafusal muscle fibers. However, the sharp
development in tension may be sufficient to excite the spindle's
sensory endings. This possibility has been used by Brown, Goodwin,
and Matthews [32] to explain the controversial "acceleration response" in the primary ending of the cat spindle. (For examples of
the response and the controversy surrounding it, see references [32,
92,114,151].) Brown et al. [32] noticed that the stimulation of
fusimotor axons would produce, or enhance, the short-lasting "acceleration response" at the onset of a ramp extension of the muscle
(Fig. 26). Furthermore, if the fusimotor nerves were stimulated
and the muscle was shortened and then restretched past the original
length, the burst of impulses called an "acceleration response"
occurred as the muscle passed through its original length, even
though there was no accelerative length change at that point [Fig.
26(C)]. The experiment has interesting implications since it not
only supplied evidence, though the separate effects of dynamic and
static motor nerve fibers, for the discrete motor innervation of
nuclear bag and nuclear chain muscle fibers, but it also draws
attention to the importance of the intrafusal muscle fibers in
determining the very high sensitivity of the primary ending to very
small displacements, a property that may be important in the reflex
control of muscle [123].

B. The Contraction of Intrafusal Muscle Fibers

Very little is known about the active mechanical properties of
intrafusal muscle fibers. This lack of knowledge is closely connected to the dearth of information on the electrical properties of
these fibers. In muscle spindles of the snake, which contain single
intrafusal muscle fibers only, the muscle fibers are reported to
twitch and to support a propagated action potential [68]. The
descriptions are qualitative, however, and it is not known whether
the intrafusal muscle fibers in the fast and slowly adapting varieties of spindle show functional differences.

The descriptions of contraction in intrafusal muscle fibers of the frog are not much more complete. Records of the twitch tensions supposed to originate in the large and small types of intrafusal muscle fiber have been reported, as have measurements of the shortening of these fibers during a twitch [155]. Such measurements are

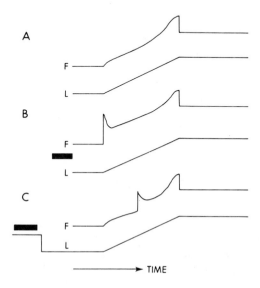

Figure 26. Diagrammatic representation of the experiment by Brown, Goodwin, and Matthews [32]. In each pair of traces F represents the instantaneous frequency of afferent impulses in the primary ending; L represents the length of the muscle. The black bar in parts B and C indicates tetanic stimulation of fusimotor axons. Ramp extension of the muscle, A, gave rise to a response in the sensory nerve that might, or might not, include an "acceleration" burst at the onset of the ramp. The burst is not shown in A. Prior stimulation of the fusimotor axons caused the initial burst to appear, as in B, or to become enhanced. In C the fusimotor axons were stimulated, the muscle shortened, and then restretched past the original length. The "acceleration" response then occurred as the muscle length passed through its initial value, even though there was no accelerative length change at that point.

both difficult to make and to interpret. The unusual properties of
the reticular zone, which is certainly present in the large intrafusal fibers, may cause blockage of the impulse and, in addition,
the high compliance of the reticular zone may be expected to affect
such records.

    Mammalian intrafusal muscle fibers have been thought to resemble
slow extrafusal muscle fibers on the indirect grounds of the graded
fashion in which fusimotor activity affects the sensory discharge
[108]. The first attempts to observe the contraction of intrafusal
muscle in the cat suggested, however, that this might not be so.
Boyd [21] reported that the intrafusal fibers twitched. Subsequent
work on rat muscle spindles [156] led to the observation that the
contractile properties of presumptive nuclear bag and nuclear chain
muscle fibers differed. The nuclear chain muscle fiber shortened
more rapidly than the nuclear bag muscle fiber. These observations
were consistent with earlier findings that the acceleration of the
sensory discharge in a primary ending was more rapid following tetanic stimulation of a static fusimotor axon than it was when the
dynamic fusimotor axon was similarly stimulated [47]. It was also
consistent with the finding that primary endings are more easily
"driven" through the static fusimotor system than through the
dynamic system [47,121]. While it was not possible in the experiments on rat spindles [156] to determine whether the contraction in
the nuclear chain muscle fibers was propagated, evidence was obtained
that the nuclear bag muscle fiber showed a local contraction only
in response to direct electrical stimulation. Concurrent work by
Diete-Spiff [55,56] and Boyd [25] on spindles from the dog and the
cat respectively, also showed that in isolated spindles nuclear bag
fibers contract in a slow, nonpropagated, fashion. More recently,
Boyd [26], using cat spindles, has been able to demonstrate the
difference in contractile properties in nuclear bag and nuclear
chain muscle fibers following stimulation of their motor nerves.
Again, the observation was that nuclear chain fibers shorten faster
than nuclear bag fibers but that the contraction in each case was

limited to the region beneath the end plates; there was no evidence for propagated activity. This observation has been confirmed in this laboratory using both direct and indirect stimulation of intrafusal muscle fibers in the cat (unpublished work).

Diete-Spiff [57] has presented records of the tension developed by intrafusal muscle fibers of the cat in response to direct electrical stimulation. The records show both fast and slow components of tension development. Diete-Spiff cautiously interpreted these records as possibly originating in one large intrafusal fiber or several small ones. Koles and Smith (unpublished work) repeated Diete-Spiff's experiment while observing the contractions of the intrafusal muscle fibers. It was found that the fast components of the tension records were associated with contraction of the nuclear chain fibers while the slow components were probably a composite record of "fused" contractions from both types of muscle fiber (Fig. 27).

The consensus of evidence is that in mammalian intrafusal muscle fibers both the nuclear bag and the nuclear chain fibers show only localized, nonpropagated contraction in response to both direct and indirect stimulation. The nuclear chain fiber, however, shortens and develops tension more rapidly than does the nuclear bag fiber. These observations are not consistent with some of the records of the electrical activity of mammalian intrafusal fibers, which suggest that one species of intrafusal muscle fiber may propagate an action potential. The electrical evidence is discussed in the next section of this review. In attempting to assess the various lines of experimental evidence it must be kept in mind that work on isolated mammalian spindles has only recently become possible. It is very difficult to know whether the isolated preparation is in a state comparable to that in the undissected muscle which has a normal blood supply. Mammalian intrafusal muscle fibers have many peculiarities, one of them may be a rapid loss of the ability to propagate an impulse when kept in vitro.

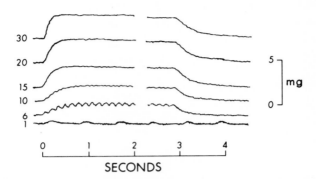

Figure 27. Records of the tension produced by a single
isolated muscle spindle from the cat lumbrical muscle in response
to direct stimulation of the intrafusal muscle fibers. The numbers
at the left side of each trace indicate the stimulus frequency in
stimuli/sec. Supra-threshold stimuli at 1 stimulus/sec gave rise
to small discrete rises in tension; nuclear chain fibers were contracting. At 6 stimuli/sec, the nuclear bag fibers were also seen
to begin to contract slowly. The records at higher stimulus rates
are a composite of tension produced by the slowly contracting nuclear bag fibers and "fused" tension developed by the more rapidly
contracting nuclear chain fibers.

VII. ELECTRICAL PROPERTIES

The electrical properties of intrafusal muscle fibers have, as
have their mechanical properties, been studied with greatest success
in the frog and the toad [33,63,65,100,105,106,154,155]. Both kinds
of intrafusal muscle fiber in the amphibia are reported to propagate
action potentials [100,154]. The muscle fibers in the two different
kinds of snake spindle also propagate an impulse [68]. The differences in electrical activity between the various structural forms
of intrafusal muscle fibers in the lower vertebrates have, however,
been given very little attention. In the toad Xenopus, it has been
found that pairs of stimuli delivered via the fusimotor nerve fibers

to the large intrafusal muscle fibers gave rise, if the stimulus
interval were greater than about 20 msec, to two impulses which
propagated through the equatorial region [154]. If, however, the
stimulus interval was decreased to about 15 msec, the second propagated response travelled only as far as the equatorial region. The
blockade of the second impulse was attributed partly to the presence
of the reticular zone in these fibers which, because of its increased membrane area per unit length of fiber, would present the
advancing impulse with an increased electrical load. Thus, the
"safety factor" for propagation would be decreased at the reticular
zone, and one assumes that during the period of altered excitability
following an impulse this low safety factor could cause blockage of
a second impulse. Paired stimuli delivered to the small type of
intrafusal muscle fiber via its motor nerve gave rise to somewhat
different results--the second impulse failed to propagate when the
stimuli were separated by as much as 200 msec [154]. In this case
the block apparently did not occur in the equatorial zone. On the
basis of this evidence, one cannot decide whether the extreme
sensitivity of the small intrafusal muscle fiber to repetitive
stimulation is caused by unusual membrane properties, or whether
the neuromuscular transmission is to be held accountable. Experiments in which the intrafusal muscle fibers are stimulated directly
would enable one to decide between these two possibilities. Preliminary experiments of this kind (Ovalle, unpublished) indicate that
the membrane properties of the small intrafusal fiber are in fact
responsible for the effect.

The sole study [106] on the nature of neuromuscular transmission
in amphibian intrafusal muscle fibers was, unfortunately, done without taking into account the fact that both the large and small motor
systems innervate the frog spindle. Thus, apart from the authors'
conclusion that the transmission is cholinergic (hardly surprising
in view of the innervation of motor units and intrafusal fibers by
branches of the same axon), no further interpretation of the evidence
can be made.

The task of investigating neuromuscular transmission in intrafusal muscle fibers has yet to be seriously undertaken. Such an
investigation may be expected to yield information which would be
valuable for the understanding of the function of the muscle spindle
in the control of skeletal muscle. The experimental difficulty of
this kind of study may not be too great since it should be possible,
for instance, to obtain very good records of miniature end-plate
potentials from such small muscle fibers [103]. As a matter of
historical interest, some of the first records of miniature end-
plate potentials were obtained from intrafusal muscle fibers in the
frog (Katz, personal communication). A necessary complement to any
such study will be the more difficult task of determining the passive membrane properties of intrafusal fibers. Values for the
membrane properties of extrafusal muscle fibers may be used to
estimate that the space constant of intrafusal fibers should be of
the order of a millimeter. One then wonders whether intrafusal
muscle fibers which are innervated at one pole only [93] can be
effectively depolarized if the muscle fiber does not propagate an
impulse.

There is very little direct evidence that intrafusal muscle
fibers in the mammal do propagate an impulse. The negative evidence,
gathered from the observation of isolated mammalian muscle spindles,
has been considered here in the section on mechanical properties.
These observations indicate that while the shortening and development of tension in nuclear chain fibers is faster than in nuclear
bag fibers, neither fiber can be shown to propagate an impulse. On
the other hand, there is a considerable amount of evidence which
suggests that if the muscle spindle is not isolated, some intrafusal
muscle fibers may propagate an impulse. Eyzaguirre [65] made the
first observation of this kind. He recorded action currents from
the surface of cat tenuissimus muscles while stimulating single
fusimotor axons. The shape of the action currents, and the fact
that curare caused them to disappear in an all-or-nothing fashion,
led him to conclude that the muscle response was propagated. Bessou

and Laporte [19] repeated the experiment but were able to divide the action currents into two kinds: a response obtained on stimulating static fusimotor axons which appeared to represent propagated activity and a response whose shape and time course suggested nonpropagated activity. The nonpropagated activity was elicited by stimulating dynamic fusimotor axons. The complex geometry and innervation of mammalian spindles does not, however, allow unequivocal interpretation of such experiments. Recently, Bessou and Pages [20] have reported that action potentials could be recorded from intrafusal muscle fibers in the cat's tenuissimus muscle when static fusimotor axons were stimulated. In these experiments the blood supply to the tenuissimus muscle was preserved. Although the muscle fibers surrounding the muscle spindle were removed so that the intrafusal muscle fibers could be seen, Bessou and Pages were unable to say whether nuclear bag or nuclear chain muscle fibers had been impaled by their microelectrode. The lack of a direct identification of the muscle fiber is unfortunate, but the total evidence from these experiments does very strongly suggest that an intrafusal muscle fiber which receives its motor innervation from static fusimotor axons is capable of propagating an impulse.

An additional finding that may be related to the ultrastructural features of intrafusal fibers is that nuclear chain fibers tend to contract as a group [25]. The close appositions, described earlier [41], between nuclear chain fibers could have significance in this regard. Whether the link between nuclear chain fibers is purely mechanical or is electrical could certainly be determined by the use of microelectrode techniques.

In summary then, the total evidence which is related to the active properties of mammalian intrafusal muscle fibers is as follows.

1. The fine structures of the fibers indicate that the nuclear chain fiber is likely to be a "fast" fiber and the nuclear bag fiber is likely to be "slow." (This is an argument based on analogy with the known structures and functions of extrafusal muscle fibers.)

2. Functional evidence suggests that nuclear chain fibers are innervated by static fusimotor axons while nuclear bag fibers are innervated separately by dynamic fusimotor axons. (Here we note that the morphological evidence does not entirely support this view.)

3. Nuclear chain fibers, as seen in the isolated muscle spindle, have a faster mechanical activity than the nuclear bag fibers (although neither nuclear bag nor nuclear chain fibers have been shown to propagate an impulse when the spindle is isolated).

4. In the unisolated spindle indirect evidence suggests that those fibers innervated by static fusimotor axons contract faster than the fibers innervated by dynamic fusimotor axons.

5. Records of the electrical activity of intrafusal muscle fibers indicate that the static fusimotor system gives rise to propagated impulses while the dynamic fusimotor system does not.

Thus, the most reasonable hypothesis to adopt at the present time is that the nuclear chain fiber is a "fast" fiber with all the properties that this term implies. It propagates an electrical impulse and consequently develops a twitch-type contraction. All evidence indicates that the nuclear bag fiber does not propagate an impulse and has a slow, graded contraction. One immediately wonders why evidence for propagated activity in nuclear chain fibers has not been directly obtained from experiments with the isolated spindle. Here the histochemical characteristics of mammalian intrafusal fibers may point to the answer. The nuclear chain fibers display a very high level of oxidative enzymes; the activity of such enzymes is often much greater than in any type of extrafusal muscle fiber. Consequently, one supposes that the function of the nuclear chain fiber is very dependent on the integrity of its oxidative metabolism. Separation of the spindle from its normal blood supply, and the inadequacy of the currently used procedures in maintaining isolated spindles in vitro may explain the discrepancy between experiments conducted in the animal and those conducted in a dish.

VIII. CONCLUSION

Everything we have had to say here about the structure and function of intrafusal muscle fibers indicates that these fibers are specialized in numerous ways, making them distinct from the extrafusal muscle fibers of the same muscle. This is true even though the intra- and extrafusal muscle may, in some cases, share motor innervation from the same motoneuron. Their development and innervation patterns, the changes in the fibers attendant on denervation and disuse, and their histochemical characteristics, all point toward multiple trophic influences from both sensory and motor nerves as having a role in determining the known specializations in structure and function. Thus, it is apparent that the extension of general principles, which arise from the study of extrafusal muscle, to predict properties of intrafusal muscle may not always be successful. Greater burdens are then placed on the investigator: a greater variety of hypotheses may be entertained and, consequently, the experimental tests need to be as direct as possible. This review indicates that reliable direct observations of the properties of intrafusal muscle fibers may be very difficult to achieve.

Ultrastructural work on intrafusal fibers has shown a great increase in the past few years. The work that has appeared has led to a better definition of the morphological types of fiber. This is particularly true in the mammal where the nuclear chain and nuclear bag fibers have structural characteristics that parallel very satisfactorily the observed functional characteristics. In addition, the use of ultrastructural and histochemical techniques has led to the finding that spindles in the rabbit contain more than one type of intrafusal fiber. The assumption that rabbit spindles contained only one type of intrafusal muscle fiber has recently (mis)directed a number of functional studies. Despite these successes, ultrastructural work on intrafusal fibers is still in its infancy. The problems raised have equalled those solved,

and the areas of development and experimental modification of the
spindle have received very little, or no, attention.

Recently, very precise formulations of the "mechanical hypothesis" on the origin of rate sensitivity in the afferent discharge
have appeared. Several attempts have been made to test the hypothesis in amphibian spindles. The results indicate that the mechanical
properties of these intrafusal fibers are not as predicted by the
hypothesis, although the notion may yet be found to be applicable
to mammalian spindles. In the future one may expect to see these
ideas tested more adequately. At the same time, interest is likely
to swing toward the small-displacement sensitivity of muscle
spindles and its possible origin in short-range elastic forces in
the intrafusal fibers.

It is obvious that the properties of intrafusal muscle fibers
impose restrictions on current ideas on the role the muscle spindle
plays in the control of movement. For example, the evidence, considered above, that the mechanical properties of intrafusal fibers
in the frog do not allow the passive spindle to act as a stretch
receptor would seem to exclude the spindle from any role in the
initiation of movement. One has to assume then that frog spindles
at least play their major role in the correction of movements which
are in progress. In the mammals the very slow contraction of the
nuclear bag fiber must impose a large delay in the pathway: fusimotor neuron -- intrafusal muscle fiber -- sensory nerve fiber --
alpha motor neuron. Hence, it seems likely that rapid changes in
this neural loop would be better accommodated by using the nuclear
chain fiber via the appropriate fusimotor axon. It is not our
intention to speculate in any depth along these lines; what we do
contend is that the eventual understanding of the motor control
system must include an understanding of the structure and function
of intrafusal muscle fibers.

ACKNOWLEDGMENTS

The writing of this review, and some of the work referred to, was supported by the Medical Research Council of Canada and the Muscular Dystrophy Association of Canada.

REFERENCES

[1] Adal, M., "The fine structure of the sensory region of cat muscle spindles," J. Ultrastruct. Res., 26, 332-354 (1969).

[2] Adal, M. N., and D. Barker, "The fine structure of cat fusimotor endings," J. Physiol. (London), 192, 50-52 (1967).

[3] Adams, R. D., D. Denny-Brown, and C. M. Pearson, Diseases of Muscle, A Study in Pathology, 2nd ed., Harper, New York, 1962.

[4] Andersson-Cedergren, E., and U. Karlsson, "Polyribosomal organization in intact intrafusal muscle fibres," J. Ultrastruct. Res., 19, 409-416 (1967).

[5] Andrew, B. L., Control and Innervation of Skeletal Muscle, D. C. Thomson, Dundee, 1966.

[6] Angers, D., "Modèle méchanique de fuseau neuromusculaire déefferenté: terminaisons primaires et secondaires," Compt. Rend. Acad. Sci. (Paris), 261, 2255-2258 (1965).

[7] Barker, D., "Some results of a quantitative histological investigation of stretch receptors in limb muscles of the cat," J. Physiol. (London), 149, 7-9 (1959).

[8] Barker, D., Symposium on Muscle Receptors, Hong Kong University Press, Hong Kong, 1962.

[9] Barker, D., "The structure and distribution of muscle receptors," in Symposium on Muscle Receptors (D. Barker, ed.), Hong Kong University Press, Hong Kong, 1962, pp. 227-240.

[10]   Barker, D., "Three types of motor ending in cat spindles,"
J. Physiol. (London), 186, 27-28 (1966).

[11]   Barker, D., "L'innervation motrice du muscle strié des
vertébrés," Actual. Neurophysiol., 8, 23-71 (1968).

[12]   Barker, D., "Fusimotor innervation," in Research Concepts in
Muscle Development and the Muscle Spindle (B. Banker, R. Przybylski,
J. Van Der Muelen, and M. Victor, eds.), Excerpta Medica, Amsterdam
(in press).

[13]   Barker, D., and M. Cope, "Tandem muscle spindles in the frog,"
J. Anat., 96, 49-57 (1962).

[14]   Barker, D., and J. L. Gidumal, "The morphology of intrafusal
muscle fibres in the cat," J. Physiol. (London), 157,513-528 (1961).

[15]   Barker, D, and M. C. Ip, "Sprouting and degeneration of
mammalian motor axons in normal and deafferented skeletal muscle,"
Proc. Roy. Soc., B.163, 538-554 (1966).

[16]   Barker, D., and M. J. Stacey, "Rabbit intrafusal muscle
fibres," J. Physiol. (London), 210, 70-72 (1970).

[17]   Barker, D., M. J. Stacey, M. N. Adal, "Fusimotor innervation
in the cat," Phil. Trans. Roy. Soc. (London), B.258, 315-346 (1970).

[18]   Batten, F., "The muscle spindle under pathological conditions,"
Brain, 20, 138-179 (1897).

[19]   Bessou, P. and Y. Laporte, "Potentials fusoraeux provoqués
par la stimulation des fibres fusimotrices chez le chat," Compt.
Rend. Acad. Sci. (Paris), 260, 4827-4830 (1965).

[20]   Bessou, P., and B. Pages, "Intracellular recording from
spindle muscle fibres of potentials elicited by static fusimotor
axons in the cat," Life Sci., 8, 417-419 (1969).

[21]   Boyd, I. A., "An isolated muscle spindle preparation," J.
Physiol. (London), 144, 11-12 (1958).

[22]   Boyd, I. A., "The diameter and distribution of the nuclear

bag and nuclear chain muscle fibres in the muscle spindles of the cat," J. Physiol. (London), 153, 23-24 (1960).

[23] Boyd, I. A., "The nuclear-bag fibre and nuclear-chain fibre system in the muscle spindles of the cat," in Symposium on Muscle Receptors, (D. Barker, ed.), Hong Kong University Press, Hong Kong, 1962, pp. 185-190.

[24] Boyd, I. A., "The structure and innervation of the nuclear bag muscle fibre system and the nuclear chain muscle fibre system in mammalian muscle spindles," Phil. Trans. Roy. Soc. (London), B. 245, 81-136 (1962).

[25] Boyd, I. A., "The behaviour of isolated mammalian muscle spindles with intact innervation," J. Physiol. (London), 186, 109-110 (1966).

[26] Boyd, I. A., "The mechanical properties of mammalian intra-fusal muscle fibres," J. Physiol. (London), 187, 10-12 (1966).

[27] Boyd, I. A., and M. R. Davey, "The distribution of two types of small motor nerve fibre to different muscles in the hind limb of the cat," in Nobel Symposium I. Muscular Afferents and Motor Control (R. Granit, ed.), Almqvist and Wiksell, Stockholm, 1966, pp. 59-68.

[28] Bravo-Rey, M. C., J. N. Yamasaki, E. Eldred, and A. Maier, "Ionizing irradiation on development of the muscle spindle," Exptl. Neurol., 25, 595-602 (1969).

[29] Bridgman, C., E. Eldred, and B. Eldred, "The distribution and structure of muscle spindles in the extensor digitorum brevis of the cat," Anat. Rec., 143, 219-227 (1962).

[30] Bridgman, C. F., E. E. Shumpert, and E. Eldred, "Insertions of intrafusal fibres in muscle spindles of the cat and other mammals," Anat. Rec., 164, 391-402 (1969).

[31] Brown, M. C., "The effect of suxamethonium on the spindles of frog sartorius and iliofibularis muscles," J. Physiol. (London), 206, 27-28 (1970).

[32]   Brown, M. C., G. M. Goodwin, and P. B. C. Matthews, "After-effects of fusimotor stimulation on the response of muscle spindle primary afferent endings," J. Physiol. (London), 205, 677-694 (1969).

[33]   Buchthal, F., and V. Jahn, "Spontaneous activity in isolated muscle spindles," Acta Physiol. Scand., 42 Suppl., 145, 25-26 (1957).

[34]   Cazzato, G., and J. N. Walton, "The pathology of the muscle spindle. A study of biopsy material in various muscular and neuro-muscular diseases," J. Neurol. Sci., 7, 15-70 (1968).

[35]   Cheng, K., and G. M. Breinin, "A comparison of the fine structure of extraocular and interosseous muscles in the monkey," Invest. Ophthalmol., 5, 535-549 (1966).

[36]   Coërs, C., "Histochemical identification of motor nerve," endings in muscle spindles," in Symposium on Muscle Receptors (D. Barker, ed.), Hong Kong University Press, Hong Kong, 1962, pp. 221-226.

[37]   Coërs, C., and J. Durand, "Donnees morphologiques nouvelles sur l'innervation des faseaux neuromusculaires," Arch. Biol. (Liege), 67, 685-715 (1956).

[38]   Cooper, S., "Muscle spindles and other muscle receptors," in Structure and Function of Muscle (G. H. Bourne, ed.), Vol. 1, Chap. XI, Academic Press, New York, 1960, pp. 381-420.

[39]   Cooper, S., and P. M. Daniel, "Human muscle spindles," J. Physiol. (London), 133, 1 (1956).

[40]   Cooper, S., and P. M. Daniel, "Muscle spindles in man; their morphology in the lumbricals and the deep muscles of the neck," Brain, 86, 563-586 (1963).

[41]   Corvaja, N., V. Marinozzi, and O. Pompeiano, "Close appositions and junctions of plasma membranes of intrafusal fibres in mammalian muscle spindles," Pflugers Arch., 296, 337-347 (1967).

[42]   Corvaja, N., V. Marinozzi, and O. Pompeiano, "Muscle spindles in the lumbrical muscle of the adult cat. Electron microscopic

observation and functional considerations," Arch. Ital. Biol., 107, 365-543 (1969).

[43] Corvaja, N., and O. Pompeiano, "The differentiation of two types of intrafusal muscle fibres in rabbit muscle spindles," Pflugers Arch., 317, 187-197 (1970).

[44] Couteaux, R., "Recherches sur l'histogenése du muscle strié et de la formation des plaques motrices," Bull. Biol., 75, 101-239 (1941).

[45] Crowe, A., "A mechanical model of the mammalian muscle spindle," J. Theoret. Biol., 21, 21-41 (1968).

[46] Crowe, A., "A mechanical model of muscle and its application to the intrafusal fibres of the mammalian muscle spindle," J. Biomechanics, 3, 583-592 (1970).

[47] Crowe, A., and P. B. C. Matthews, "Further studies of static and dynamic fusimotor fibres," J. Physiol. (London), 174, 132-151 (1964).

[48] Crowe, A., and A. Ragab, "Studies on the fine structure of the capsular region of tortoise muscle spindles," J. Anat., 107, 257-269 (1970).

[49] Cuajunco, F., "Embryology of the neuromuscular spindles," Contrib. Embryol. Carnegie Inst., 19, (1927).

[50] Cuajunco, F., "Development of the neuromuscular spindle in human fetuses," Contrib. Embryol. Carnegie Inst., 28, 95-128 (1940).

[51] Daniel, P. M., and S. J. Strich, "Abnormalities in the muscle spindles in dystrophia myotonica," Neurol., 14, 310-316 (1964).

[52] De Anda, G., and M. A. Rebollo, "Histochemistry of the neuro-muscular spindles in the chicken during development," Acta Histo-chem., 31, 287-295 (1968).

[53] De Reuck, A., and J. Knight, Myotatic, Kinesthetic and Vestibular Mechanisms, J. and A. Churchill, Ciba Foundation Symposium, London, 1967.

[54] Dickson, L. M., "The development of nerve-endings in the
respiratory muscles of the sheep," J. Anat., 74, 268-276 (1940).

[55] Diete-Spiff, K., "Slow contraction of intrafusal muscle fibres
of lumbrical muscle spindles of the dog," J. Physiol. (London), 183,
65-66 (1966).

[56] Diete-Spiff, K., "Time course of mammalian intrafusal muscle
contraction as revealed by cine photography of isolated lumbrical
muscle spindles of the dog," Arch. Ital. Biol., 104,354-386 (1966).

[57] Diete-Spiff, K., "Tension development by isolated muscle
spindles in the cat," J. Physiol. (London), 193, 31-43 (1967).

[58] Dubowitz, V., and A. Pearse, "A comparative histochemical
study of oxidative enzyme and phosphorylase activity in skeletal
muscle," Histochemie., 2, 105-117 (1960).

[59] During, M., and K. H. Andres, "Zur Feinstruktur der Muskelspindel von Mammalia," Anat. Anz., 124, 566-573 (1969).

[60] Eldred, E., C. Bridgman, J. Swett, and B. Eldred, "Quantitative
comparisons of muscle receptors of the cat's medial gastrocnemius,
soleus and extensor digitorum brevis muscles," in Symposium on
Muscle Receptors, (D. Barker, ed.), Hong Kong University Press,
Hong Kong, 1962, pp. 207-213.

[61] Eldred, E., H. Yellin, L. Gabbois, and S. Sweeney, "Bibliog-
raphy on muscle receptros; their morphology, pathology, and
physiology," Exp. Neurol., Suppl. 3 (1967).

[62] Engel, W. K., and R. L. Irwin, "A histochemical-physiological
correlation of frog skeletal muscle fibres," Am. J. Physiol., 213,
511-518 (1967).

[63] Eyzaguirre, C., "Functional organization of neuromuscular
spindle in the toad," J. Neurophysiol., 20, 523-542 (1957).

[64] Eyzaguirre, C., "Modulation of sensory discharges by efferent
spindle excitation," J. Neurophysiol., 21, 465-480 (1958).

[65]   Eyzaguirre, C., "The electrical activity of mammalian intra-fusal fibres," J. Physiol. (London), 150, 169-185 (1960).

[66]   Farrell, P. R., and M. R. Fedde, "Uniformity of structural characteristics throughout the length of skeletal muscle-fibres," Anat. Rec., 164, 219-230 (1969).

[67]   Forster, L., "Zur Kenntniss der Muskelspindeln," Virchows Arch. Pathol. Anat., 137, 121-154 (1894).

[68]   Fukami, Y., "Tonic and phasic muscle spindles in snake," J. Neurophysiol., 33, 28-45 (1970).

[69]   Fukami, Y., and C. C. Hunt, "Structure of snake muscle spindles," J. Neurophysiol., 33, 9-27 (1970).

[70]   Gauthier, G. F., and H. A. Padykula, "Cytological studies of fibre types in skeletal muscle: A comparative study of the mam-malian diaphragm," J. Cell Biol., 28, 333-354 (1966).

[71]   Germino, N. I., and H. D'Albora, "Succinic dehydrogenase activity in the neuromuscular spindles of the chick," Experientia, 21, 45-46 (1965).

[72]   Gottlieb, G. L., G. C. Agarwal, and L. Stark, "Stretch receptor models. 1. Single-efferent single-afferent innervation," I.E.E.E. Trans. Man-Machine Systems, 10, 17-27 (1969).

[73]   Granit, R., Muscular Afferents and Motor Control, Almqvist and Wiksell, Nobel Symposium I. Stockholm, 1966.

[74]   Granit, R., The Basis of Motor Control. Integrating the Activity of Muscles, Alpha and Gamma Motoneurons and their Leading Control Systems, Academic Press, New York, 1970.

[75]   Gray, E. G., "The spindle and extrafusal innervation of a frog muscle," Proc. R. Soc. London, Series B., 146, 416-430 (1957).

[76]   Gray, E. G., "The structures of fast and slow muscle fibres in the frog," J. Anat., 92, 559-562 (1958).

[77]   Gruner, J., "La structure fine du fuseau neuromusculaire humain," Rev. Neurol., 104, 490-507 (1961).

[78] Guth, L., and F. J. Samaha, "Qualitative differences between actomyosin ATPase of slow and fast mammalian muscle," Exper. Neurol., 25, 138-152 (1969).

[79] Guth, L., F. J. Samaha, and R. W. Albers, "The neural regulation of some phenotypic differences between the fibre types of mammalian skeletal muscle," Exper. Neurol., 26, 126-135 (1970).

[80] Gutmann, E., and J. Zelena, "Morphological changes in the denervated muscle," in The Denervated Muscle (E. Gutmann, ed.), Publishing House of the Czechoslovak Academy of Sciences, Prague, 1962, pp. 57-102.

[81] Haggqvist, G., "A study of the histology and histochemistry of the muscle spindles," Z. Biol., 112, 11-26 (1960).

[82] Henneman, E., and C. B. Olson, "Relations between structure and function in design of skeletal muscle," J. Neurophysiol., 28, 581-598 (1965).

[83] Hennig, G., "Die Nervenendigungen der Rattenmuskelspindel im elektronen und phasenkontrastmikroskopischen Bild," Z. Zellforsch. Mikrosk. Anat., 96, 275-294 (1969).

[84] Hess, A., "Two kinds of motor nerve endings on mammalian intrafusal muscle fibres revealed by the cholinesterase technique," Anat. Rec., 139, 173-184 (1961).

[85] Hess, A., "Vertebrate slow muscle fibres," Physiol. Rev., 50, 40-62 (1970).

[86] Hewer, E. E., "The development of nerve endings in the human foetus," J. Anat., 69, 369-379 (1935).

[87] Hill, D. K., "Tension due to interaction between the sliding filaments in resting striated muscle: The effect of stimulation," J. Physiol. (London), 199, 637-684 (1968).

[88] Houk, J. C., "Rate sensitivity of mechanoreceptors," Ann. N.Y. Acad. Sci., 156, 901-916 (1969).

[89] Houk, J., R. W. Cornew, and L. Stark, "A model of adaptation in amphibian spindle receptors," J. Theoret. Biol., 12, 196-215 (1966).

[90] Hubbard, S. J., and A. Hess, "The fine structure of the primary sensory zone of cat muscle spindles," Anat. Rec., 157, 262 (abstr.) (1967).

[91] James, N. T., "Histochemical demonstration of myoglobin in skeletal muscle fibres and muscle spindles," Nature, 219, 1174-1175 (1968).

[92] Jansen, J. K. S., and P. B. C. Matthews, "The central control of the dynamic response of muscle spindle receptors," J. Physiol. (London), 161, 357-378 (1962).

[93] Jones, E. G., "The innervation of muscle spindles in the Australian opposum, Trichosurus valpecula, with special reference to the motor nerve endings," J. Anat., 100, 733-759 (1966).

[94] Kalugina, M. A., "Development of receptors in skeletal muscles of mammals," (Russian), Arch. Anat. Gistol. Embriol., 33, 59-63 (1956).

[95] Karlsson, U. L., "Differentiated zonal stretch of frog intra- fusals," in Research Concepts in Muscle Development and the Muscle Spindle, (B. Banker, R. Przybylski, J. Van Der Meulen, and M. Victor, eds.), Excerpta Medica, Amsterdam, (in press).

[96] Karlsson, U., and E. Andersson-Cedergren, "Motor myoneural junctions on frog intrafusal muscle fibres," J. Ultrastruct. Res., 14, 191-211 (1966).

[97] Karlsson, U., and E. Andersson-Cedergren, "Small leptomeric organelles in intrafusal muscle fibres of the frog as revealed by electron microscopy," J. Ultrastruct. Res., 23, 417-426 (1968).

[98] Karlsson, U., E. Andersson-Cedergren, and D. Ottoson, "Cellular organization of the frog muscle spindle as revealed by serial sections for electron microscopy," J. Ultrastruct. Res., 14, 1-35 (1966).

[99] Karpati, G., and W. K. Engel, "Transformation of the histochemical profile of skeletal muscle by 'foreign' innervation,"
Nature, 215, 1509-1510 (1967).

[100] Katz, B., "The efferent regulation of the muscle spindle in
the frog," J. Exp. Biol., 26, 201-217 (1949).

[101] Katz, B., "Depolarization of sensory terminals and the
initiation of impulses in the muscle spindle," J. Physiol. (London),
111, 248-260 (1950).

[102] Katz, B., "The terminations of the afferent nerve fibre in
the muscle spindle of the frog," Proc. R. Soc. London, Series B.,
243, 221-240 (1961).

[103] Katz, B., and S. Thesleff, "On the factors which determined
the amplitude of the 'miniature end-plate potential,'" J. Physiol.
(London), 137, 267-278 (1957).

[104] Kelly, A. M., and S. I. Zachs, "The histogenesis of rat
intercostal muscle," J. Cell Biol., 42, 135-153 (1969).

[105] Koketsu, K., and S. Nishi, "Action potentials of single
intrafusal muscle fibres of frogs," J. Physiol. (London), 137, 193-
209 (1957).

[106] Koketsu, K., and S. Nishi, "An analysis of junctional potentials of intrafusal muscle fibres in frogs," J. Physiol. (London),
139, 15-26 (1957).

[107] Kruger, P., and P. G. Gunther, "Fasern mit 'Fibrillenstruktur'
und Fasern mit 'Felderstruktur' in der quergestreiften Skeletmus-
kulatur der Sauger and des Menschen," Z. Ges. Anat., 118, 313-323
(1955).

[108] Kuffler, S. W., C. C. Hunt, and J. P. Quilliam, "Function of
medullated small-nerve fibres in mammalian ventral roots: efferent
muscle spindle innervation," J. Neurophysiol., 14, 29-54 (1951).

[109] Landon, D. N., "Electron microscopy of muscle spindles," in
Control and Innervation of Skeletal Muscle, (B. L. Andrew, ed.),
D. C. Thomson, Dundee, 1966, pp. 96-111.

[110]   Langworthy, O. R., "A study of the innervation of the tongue musculature with particular reference to the proprioceptive mechanism," J. Comp. Neurol., 36, 273-297 (1924).

[111]   Lannergren, J., and R. S. Smith, "Types of muscle fibres in toad skeletal muscle," Acta Physiol. Scand., 68, 263-274 (1966).

[112]   Lapresle, J., and M. Milhaud, "Pathologie du fuseau neuro-musculaire," Rev. Neurol., 110, 97-112 (1964).

[113]   Latyshev, V. A., "New formation processes in skeletal musculature of man and mammals" (Russian), Uch. Zap. Krans. Gos. Ped. Inst., 14, (1958). (Quoted from Intern. Abstr. Biol. Sci., 11, Abstr. 1742.)

[114]   Lennerstrand, G., and U. Thoden, "Dynamic analysis of muscle spindle endings in the cat using length changes of different length-time relations," Acta Physiol. Scand., 73, 234-250 (1968).

[115]   Lippold, O. C. J., J. G. Nicholls, and J. W. T. Redfearn, "Electrical and mechanical factors in the adaptation of a mammalian muscle spindle," J. Physiol. (London), 153, 209-217 (1960).

[116]   Liu, H. C., and R. M. Maneely, "The development of muscle spindles in the embryonic and regenerative tail of Hemidactylus Bowringi (Gray)," Acta Anat., 72, 63-74 (1969).

[117]   Marchand, E. R., and E. Eldred, "Postnatal increase of intrafusal fibres in the rat muscle spindle," Exper. Neurol., 25, 655-676 (1969).

[118]   Mark, R. F., L. R. Marotte, and J. R. Johnstone, "Reinnervated eye muscles do not respond to impulses in foreign nerves," Science, 170, 193-194 (1970).

[119]   Matthews, B. H. C., "The response of a single end organ," J. Physiol. (London), 71, 64-110 (1931).

[120]   Matthews, B. H. C., "Nerve endings in mammalian muscle," J. Physiol. (London), 78, 1-53 (1933).

[121]   Matthews, P. B. C., "The differentiation of two types of fusimotor fiber by their effects on the dynamic response of muscle

spindle primary endings," Quart. J. Exp. Physiol., 47, 324-333
(1962).

[122] Matthews, P. B. C., "Muscle spindles and their motor control,"
Physiol. Rev., 44, 220-286 (1964).

[123] Matthews, P. B. C., "Evidence that the secondary as well as
the primary endings of the muscle spindle may be responsible for
the tonic stretch reflex of the decerebrate cat," J. Physiol.
(London), 204, 365-393 (1969).

[124] Matthews, P. B. C., and D. R. Westbury, "Some effects of
fast and slow motor fibres on muscle spindle of the frog," J. Physiol.
(London), 178, 178-192 (1965).

[125] Mayr, R., "Untersuchungen an isolieten Muskelspindeln der
Ratte nach Cholinesterasedarstellung und Sudanschwarz-Farbung,"
Z. Zellforsch., 93, 594-606 (1969).

[126] Mayr, R., "Zwei elektronenmikroskopisch unterscheidbare
Formen sekundärer sensorischer Endigungen in einer Muskelspindel
der Ratte," Z. Zellforsch, 110, 97-107 (1970).

[127] Meier, H., "The muscle spindle in mice with hereditary
neuromuscular diseases," Experientia, 25, 965-968 (1969).

[128] Merrillees, N. C. R., "The fine structure of muscle spindles
in the lumbrical muscles of the rat," J. Biophys. Biochem. Cytol.,
7, 725-742 (1960).

[129] Merton, P. A., "Speculations on the servo control of movement," in The Spinal Cord, Ciba Symposium, Churchill, London, 1953,
pp. 247-260.

[130] Mummenthaler, M., and W. K. Engel, "Cytological localization
of cholinesterase in developing chick embryo skeletal muscle," Acta
Anat., 47, 274-299 (1961).

[131] Nakajima, Y., T. R. Shantha, and G. H. Bourne, "Enzyme
histochemical studies on the muscle spindle," Histochemie, 16, 1-8
(1968).

[132] Nyström, B., "Muscle-spindle histochemistry," Science, 155, 1424-1426 (1967).

[133] Ogata, T., and M. Mori, "Histochemical study of oxidative enzymes in vertebrate muscles," J. Histochem. Cytochem., 12, 171-182 (1964).

[134] Ottoson, D., and G. M. Shepherd, "Length changes within isolated frog muscle spindle during and after stretching," J. Physiol. (London), 207, 747-760 (1970).

[135] Ottoson, D., and G. M. Shepherd, "Steps in impulse generation in the isolated muscle spindle," Acta Physiol. Scand., 79, 420-423 (1970).

[136] Ovalle, W. K., "Fine structure of rat intrafusal muscle fibers. The polar region," J. Cell Biol., 51, 83-103 (1971).

[137] Ovalle, W. K., "Morphology and distribution of motor nerve terminals on rat intrafusal muscle fibers," Anat. Rec., 172, 378 (abstr.) (1972).

[138] Page, S. G., "A comparison of the fine structures of frog slow and twitch muscle fibres," J. Cell Biol., 26, 477-497 (1965).

[139] Page, S. G., "Intrafusal muscle fibres in the frog," J. Microscopie, 5, 101-104 (1966).

[140] Patel, A. N., V. S. Lalitha, and D. K. Dastur, "The spindle in normal and pathological muscle: An assessment of the histological changes," Brain, 91, 737-750 (1968).

[141] Payton, B. W., "Histological staining properties of procion yellow M4RS," J. Cell Biol., 45, 659-662 (1970).

[142] Porayko, O., and R. S. Smith, "Morphology of muscle spindles in the rat," Experientia, 24, 588-589 (1968).

[143] Rebollo, M. A., and G. De Anda, "The neuromuscular spindles in the adult chicken. II. Histochemistry," Acta Anat., 67, 595-608 (1967).

[144] Robertson, J. D., "Electron microscopy of the motor end-plate and the neuromuscular spindle," Am. J. Phys. Med., 39, 1-43 (1960).

[145] Rudjord, T., "A second order mechanical model of muscle spindle primary endings," Kybernetik, 6, 205-213 (1970).

[146] Rumpelt, H. J., and H. Schmalbruch, "Zur Morphologie der Bauelemente von Muskelspindeln bei Mensch und Ratte," Z. Zellforsch. Mikrosk. Anat., 102, 601-630 (1969).

[147] Samaha, F. J., L. Guth, and R. W. Albers, "Phenotypic differences between the actomyosin ATPase of the three fibre types of Mammalian skeletal muscle," Exper. Neurol., 26, 120-125 (1970).

[148] Samaha, F. J., L. Guth, and R. W. Albers, "The neural regulation of gene expression in the muscle cell," Exp. Neurol., 27, 276-282 (1970).

[149] Sampaolo, C. L., "La morfogenesi dei fusi neuromuscolari nel pollo," Quad. Anat. Prat., 21, 1-38 (1965).

[150] Scalzi, H. A., and H. M. Price, "Ultrastructure of the sensory region of the mammalian muscle spindle," J. Cell Biol., 43, 124 (abstr.) (1969).

[151] Schäfer, S. S., "The acceleration response of a primary muscle spindle ending to ramp stretch of the extrafusal muscle," Experientia, 23, 1026-1027 (1967).

[152] Schröder, J. M., "Zur pathologenese der Isoniazid Neuropathie II," Acta Neuropath., 16, 324-341 (1970).

[153] Sherrington, C. S., "On the anatomical constitution of nerves of skeletal muscles; with remarks on recurrent fibres in the ventral spinal nerve-root," J. Physiol. (London), 17, 211-258 (1894).

[154] Smith, R. S., "Activity of intrafusal muscle fibres in muscle spindles of Xenopus laevis," Acta Physiol. Scand., 60, 223-239 (1964).

[155] Smith, R. S., "Contraction in intrafusal muscle fibres of Xenopus laevis following stimulation of their motor nerves," Acta Physiol. Scand., 62, 195-208 (1964).

[156] Smith, R. S., "Properties of intrafusal muscle fibres," in Muscular Afferents and Motor Control. Nobel Symposium I, (R. Granit, ed.), Almqvist and Wiksell, Stockholm, 1966, pp. 69-80.

[157] Smith, R. S., "Sensory transduction in frog muscle spindles: role of the intrafusal muscle fibres," in Research Concepts in Muscle Development and the Muscle Spindle (B. Banker, R. Przybylski, J. Van Der Muelen, and M. Victor, eds.), Excerpta Medica, Amsterdam (in press).

[158] Spiro, A. J., and R. L. Beilin, "Histochemical duality of rabbit intrafusal fibers," J. Histochem. Cytochem., 17, 348-349 (1969).

[159] Spiro, A. J., and R. L. Beilin, "Human muscle spindle histo-chemistry," Arch. Neurol., 20, 271-275 (1969).

[160] Stein, J. M., and H. A. Padykula, "Histochemical classifica-tion of individual skeletal muscle fibers of rat," Am. J. Anat., 110, 103-124 (1962).

[161] Studifskij, A. N., and A. P. Striganova, The Restitutional Process in Skeletal Muscle, (Russian), Medgiz, Moscow, 1951.

[162] Sunderland, S., and L. J. Ray, "Denervation changes in mammalian striated muscle," J. Neurol. Neurosurg. Psychiat., 13, 159-172 (1950).

[163] Sutton, A. C., "On the development of the neuromuscular spindle in the extrinsic eye muscles of the pig," Am. J. Anat., 18, 117-144 (1915).

[164] Swett, J. E., and E. Eldred, "Comparison in structure of stretch receptors in medial gastrocnemius and soleus muscles of the cat," Anat. Rec., 137, 461-473 (1960).

[165] Tello, J. F., "Die Entehung der motorischen und sensiblen Nervenendigungen," Z. Anat. Entwick., 64, 348-440 (1922).

[166] Tower, S., "Atrophy and degeneration in the muscle spindle," Brain, 55, 77-89 (1932).

[167]   Toyama, K., "An analysis of impulse discharges from the spindle receptor," Jap. J. Physiol., 16, 113-125 (1965).

[168]   Walker, L. B., "Diameter spectrum of intrafusal muscle fibers in muscle spindles of the dog," Anat. Rec., 130, 385 (1958).

[169]   Weissmann, A., "Ueber die Verbindung der Muskelfasern mit ihren Ansatzpunkten," Z. Ration. Med., 12, 126-144 (1861).

[170]   Wirsen, C., "Histochemical heterogeneity of muscle spindle fibres," J. Histochem. Cytochem., 12, 308-309 (1964).

[171]   Wirsen, C., and K. S. Larsson, "Histochemical differentiation of skeletal muscle in foetal and newborn mice," J. Embryol. Exptl. Morph., 12, 759-767 (1964).

[172]   Yellin, H., "A histochemical study of muscle spindles and their relationship to extrafusal fiber types in the rat," Am. J. Anat., 125, 31-46 (1969).

[173]   Yellin, H., "Unique intrafusal and extraocular muscle fibers exhibiting dual actomyosin ATPase activity." Exper. Neurol., 25, 153-163 (1969).

[174] Yellin, H., and E. Eldred, "Spindle activity of the tenotomized gastrocnemius muscle in the cat," Exper. Neurol., 29, 513-533 (1970).

[175]   Zelena, J., "Morphogenetic influence of innervation on the ontogenetic development of muscle spindles," J. Embryol. Exp. Morph. 5, 283-292 (1957).

[176]   Zelena, J., "The effect of denervation on muscle development," in The Denervated Muscle (E. Gutmann, ed.), Publishing House of the Czechoslovak Academy of Sciences, Prague, 1962, pp. 103-126.

[177]   Zelena, J., "Development, degeneration, and regeneration of receptor organs," in Mechanisms of Neural Regeneration, Prog. Brain Res. Vol. 13 (M. Singer and J. P. Schade, eds.), Elsevier Publishing Co., 1964, pp. 175-211.

[178]   Zelena, J., and P. Hnik, "Effect of innervation on the development of muscle receptors," in The Effect of Use and Disuse

on Neuromuscular Functions, (E. Gutmann and P. Hnik, eds.), Publishing House of the Czechoslovak Academy of Sciences, Prague, 1963, pp. 95-105.

CHAPTER 8

MULTIPLICITY OF MUSCLE CHANGES POSTULATED
FROM MOTONEURON ABNORMALITIES

W. King Engel and John R. Warmolts

Medical Neurology Branch
National Institute of Neurological Diseases and Stroke
National Institutes of Health
Public Health Service
U.S. Department of Health, Education, and Welfare
Bethesda, Maryland

I.   INTRODUCTION

A motor unit is defined as one alpha lower motoneuron and the several hundred myofibers innervated by it. Neuromuscular diseases are disorders in which there is a major disturbance somewhere in the motor unit. The disturbance itself can originate within the motor unit or be secondary to a cause outside the motor unit.

There are two basic histochemical types of myofibers with at
least two subtypes of each, but for simplicity we refer only to the
basic types, called type I and type II myofibers [1,2]. A number of
studies indicate that all myofibers within a motor unit are of the
same histochemical type which is determined by the alpha motoneuron
[3-11]. Thus we can postulate two corresponding types of alpha
motoneurons, I and II, and even corresponding types I and II Schwann
cells [2,12,13]. We have not been able histochemically to differentiate the postulated two types of alpha motoneurons using numerous
stains, including those which differentiate myofiber types. However,
as a group all the alpha motoneurons (in the cat lumbar cord) are rich
in phosphorylase and poor in succinate dehydrogenase, in contrast
with the smaller neurons, which would include gamma motoneurons,
interneurons, and Renshaw neurons, of the anterior horn which have
the opposite characteristics, i.e., no phosphorylase and rich in
succinate dehydrogenase [14-16]. Recent animal studies have shown
that myofibers of a single motor unit are widely scattered in a
muscle and not grouped next to each other [10,11,17].

II. MOTOR UNITS AS TREES

In humans the two types of motor units, with two (or more)
subtypes of each, may be thought of simply as several kinds of
trees composing a forest. For simplicity, we discuss only the two
main groups. As analogies, the base of the tree is the motoneuron
soma, the roots the dendrites, the main trunk the axon, the branches
the axonal branches, the small terminal twigs the small terminal
axonal twigs, and the leaves the myofibers (one leaf located on each
twig and dependent on the twig for its health). These trees are
mixed together in the forest, with their crowns interlaced. The
leaves are maintained in their healthy state by a continuous flow
of sap up from the base of the tree. We must fantasize on the tree
analogy two additional phenomena: (a) that when a twig, branch, or
tree dies completely its leaves will become detached and "fall"
(atrophy) very slowly, during which time many will be "caught" and

reattached to small twiglets extending from adjacent twigs of surviving trees (collateral axonal sprouting from healthy motoneurons) and
(b) with partial decrease of the upward flow of sap (or some component of the sap) from the base the leaves can shrink (atrophy)
but will remain alive and motile, not falling off. It is proposed
that the flow of sap can be impaired in different quantitative and
qualitative ways, causing different pathological manifestations in
myofibers [12,13,18,19]. Sometimes all myofibers of one type atrophy
selectively [2] (e.g., type I or type II myofiber atrophy) or show
other changes (e.g., central core disease or type I myofiber hypotrophy with central nuclei), and sometimes only random twigs and
myofibers within each motor unit are affected (e.g., in myasthenia
gravis and perhaps in myotonic atrophy). Neuronal diseases can be
thought of as affecting whole trees, individual branches, or individual twigs.

The motoneuron ("tree") can be visualized as supplying in the
"sap" various known and proposed influences or "factors" to cells
in contact with it, especially myofibers ("leaves") and Schwann
cells, and in turn being influenced by or receiving "factors" from
them [2,12,13,18,19]. The proposed "factors" each might be chemically different or be related to different patterns or amount of
release of a fewer number of substances. The term "factor" is
simply an operational one to describe an influence. From the motoneuron to myofibers there are (a) an "excitatory factor" acetylcholine, and (b) a "trophic factor." (For this discussion, trophic
factor and trophic-type factor are considered to be separate from
acetylcholine, though conceivably they both could be in the form of
different amounts or patterns of release of acetylcholine.) The
trophic factor, which may be multiple but is discussed here as
singular from a given motoneuron, maintains myofiber health and
diameter. (c) There is also a "trophic-type factor," which maintains
distinct histochemical type of the myofibers, and it must be different from the various types of motoneurons. The natures of the
trophic and trophic-type factors are not known, but one reasonable

possibility is that they are proteins, continuously produced by the abundant ribonucleoprotein of the motoneuron soma and flowing down the axon to or into the numerous myofibers of the motor unit.   (d) The other neuronal factors to muscle, such as a proposed "inhibitory factor" that prevents a myofiber from discharging spontaneously in the form of myotonia and fibrillation, could also be exported protein.   From the motoneuron to Schwann cells there is a trophic factor.   There is also a trophic factor from Schwann cells to the motoneuron and, theoretically, partial abnormality of it could be responsible for secondary weakening of a motoneuron in any disease currently attributed to the motoneuron and certainly in all "dysschwannian" neurogenic muscular atrophies.

For primary myofiber growth, the influence of the motoneuron is not needed, at least not in tissue culture.   In cultures of muscle explants grown without neurons, myoblasts fuse to form muscle straps with central nuclei (resembling myotubes), which in turn differentiate into small myofibers with nuclei in the subsarcolemmal position and packed full of myofibrils [20,21].   These myofibers vigorously contract spontaneously [22] and can do so for as long as 2 years in culture without neurons [23].   They differ from normal adult myofibers in remaining small and not being of different histochemical types [2], suggesting that the motoneuron influence is required for the complete increase of size and for differentiation into different histochemical types of myofibers.   In the intact organism, once the myofiber has accepted the motoneuron influence and attains full maturity, its vitality becomes totally dependent on that motoneuron influence, having passed beyond a point of independent existence.   Subsequent loss of motoneuron innervation results in progressive atrophy of the muscle fiber to death, appearing as pyknotic nuclear clumps, unless it is rescued by a "lifeline" in the form of a collateral axonal sprout from a surviving motor neuron with twigs in the same region [3,24].

III. EFFECTS OF LOSS OF MOTONEURON INFLUENCE

The consequences of loss of motoneuron influence on the myofibers can be profound and varied. Total loss (complete denervation)
of influence on mature myofibers results in their atrophy and
eventual complete abolition. If a disease, such as amyotrophic
lateral sclerosis, progressive muscular atrophy, peroneal muscular
atrophy (Charcot-Marie-Tooth disease), or some polyneuropathies,
affects only a portion of the motoneuron population, surviving ones
in the region produce from near their axon tips collateral sprouts
that are capable of reinnervating adjacent myofibers which have
lost their original innervation, if their atrophy has not progressed
too far [3-5,24]. These collateral sprouts of the foster motoneuron
thus incorporate the newly adopted myofibers into their unit. The
neural sprouts provide trophic factor which allows the myofibers to
return to a normal size, composition, and function. The muscle
fibers become of the histochemical type of the adopting motoneuron
because of the proposed trophic-type factor, making, in contrast to
the normal mosaic pattern, a large group of fibers together of the
same histochemical type, so-called "type-grouping" [2-5,25]. In
this process the motoneuron increases the number and population
density of its myofibers. When such an enlarged motor unit then
dies, a large group of fibers all in the same stage of atrophy
ensues.

Theoretically, if one type of motoneuron is selectively involved in a disease, the corresponding type of myofiber would be
lost selectively, the result being a surviving predominance of the
unaffected fiber type, i.e., "type predominance" [2,25].

Other findings in denervated muscle include target and targetoid
fibers [2,3,25,26]. Although it is not yet settled whether they
represent denervated or re-innervated myofibers, we prefer the
former. Sometimes in a large type-group all fibers contain similar
target or targetoid formations [25], an observation more compatible

with the idea that the target or targetoid change is the result of
simultaneous partial denervation rather than simultaneous re-inner-
vation of those fibers. Target fibers have not yet been produced
consistently in animals.

In biopsies representing every type of denervation disease of
humans we have studied (but not in every biopsy), an occasional
myofiber undergoing necrosis, phagocytosis, or regeneration has
been found, usually in a biopsy containing at least a moderate
number of atrophic myofibers. We doubt that such necrotic fibers
necessarily indicate an additional "myopathic" pathogenesis of the
disease.

A more profound atrophy of one myofiber type ("type atrophy") is
found in animals several weeks following acute and total denerva-
tion--the type II fibers show the greater atrophy, even though both
myofiber types are equally denervated [27-29]. Therefore, type
atrophy in a human disease could reflect (a) nonselective diminution
of trophic factor from both types of motoneuron to which one type
of myofiber is more susceptible, especially type II myofibers as
analogous to total denervation of animals (a nonselective neuropathy),
(b) selective abnormality of the corresponding type of motoneuron
(a selective neuropathy), or (c) selective unreceptiveness of one
type of myofiber to neural trophic factor (selective myopathy) [12,
13]. We consider the first as most likely to explain type II atrophy
of cachexia, disuse, and corticosteroid toxicity. The same three possibilities exist to explain a finding of type hypotrophy, i.e.,
selective impairment of myofiber size increase, with or without
impairment of maturation beyond the myotube state [30,31]. The
first mechanism proposed above has been demonstrated for type II hypotrophy in animals: acute and total denervation of all rat gastrocnemius myofibers at birth results 21 days later in selective
hypotrophy of only the type II myofibers, which also retain their
central nuclei and therefore resemble myotubes [31].

When any of these several above-described histologic manifestations occur in a disease, the question of a neural abnormality

must be raised, irrespective of the electromyographic (EMG)
results--in fact, whatever the EMG changes accompanying these
histologic manifestations, one must attempt to formulate a possible
explanation of the electrical phenomena on the basis of a neural
abnormality. This is the only way to expose the various possible
pathogeneses (Table 1).

IV. THEORETICAL CONSTRUCT OF ELECTROMYOGRAPHIC MANIFESTATIONS

A theoretical construct of electromyographic manifestations
resulting from disease involving the whole or parts of motoneurones
whose affected myofibers are in the vicinity of the exploratory EMG
electrode is as follows [32,33]:

1. Loss of some motoneurones in their entirety (loss of whole
trees), would reduce the number of different motor unit potentials
appearing for a given effort. Collateral sprouting with type
grouping within surviving motor units would cause their motor unit
action potentials to be increased in amplitude and duration.

2. Reduction in the number of activated myofibers within a
motor unit, whether caused by a disease randomly (a) affecting
terminal axonal twigs, (b) impairing neuromuscular junction transmission, or (c) affecting principally the myofiber, would diminish
the amplitude and duration of the motor unit potential and also
reduce the tension generated by that motor unit. For a required
amount of effort, the weakened unit would have to discharge more
rapidly and/or more than the normal number of such weakened units
would be discharging. Such a pattern of short duration, small
amplitude, overly abundant potentials for a given amount of work is
often erroneously termed "myopathic" without further qualification.
Preferably the pattern should be given a nondiagnostic name encom-
passing its features, such as our acronym SSAP.

3. In selective type atrophy, atrophic myofibers still capable
of being activated would, by virtue of their reduced diameter,
contribute single-fiber spikes with smaller amplitude and duration

TABLE 1

Proposed Likelihood of the Neuromuscular Involvement in Various
Diseases Being Caused by Abnormality of the Lower Motor Neuron

	The major aspect of the disease	A minor aspect of the disease
Amyotrophic lateral sclerosis, Progressive muscular atrophy, Infantile spinal muscular atrophy, Juvenile proximal spinal muscular atrophy	Definite	
Spinocerebellar degenerations	Definite	
"Axonal"-degeneration peripheral neuropathies	Definite	
Segmental-demyelination peripheral neuropathies	Definite	
Type II myofiber atrophy (cachexia, disuse, cortico-steroids)	Probable	
Myasthenia gravis	Probable	
Facilitating myasthenic syndrome	Probable	
Thyrotoxic atrophy	Possible	
Hypokalemic periodic paralysis, idiopathic		Possible
Facio-scapulo-humeral "dystrophy"	Possible	Possible
Myotonia congenita	Possible	
Myotonic atrophy	Probable	
Rod disease, late-onset	Possible	
Type I myofiber hypotrophy with central nuclei	Probable	
Benign congenital hypotonia with small type I fibers	Possible	
Benign congenital hypotonia with type predominance	Probable	
Central core disease	Probable	
Rod disease, congenital	Possible	
Duchenne muscular dystrophy	None	None

resulting in a similarly smaller composite motor unit potential.
The duration of composite potentials from neighboring motor units
without the atrophic myofibers might be slightly shorter due to
passive collapse of their territorial extent, but to a lesser
degree. Since quantitative measurement of motor unit potential
duration is estimated from motor units activated on weak contraction,
and since they are probably type I units [33], we should expect mean
motor unit potential duration to be shorter in type I atrophy than
type II atrophy, which it is.

V. DISEASES OF MOTONEURON TREES

A. Twigs

A disease of the small axonal twigs is botulism poisoning,
which gives an SSAP EMG [34]. Although myasthenia gravis is manifest
as a small twig disease, we consider that to be based on an abnormality of the whole tree [18], as discussed below.

B. Branches

Diseases affecting large neuronal branches would include some
examples of the segmental demyelination type of peripheral neuropathy, where the accumulated demyelination defect involves random
large branches in the tree rather than the main trunk. The involvement of axonal branches would be secondary to the Schwann cell
defect causing the segmental demyelination.

C. Whole Trees

Diseases of the whole tree, due to disease of the base (soma)
or main trunk (axon), can (a) occur before or after myofiber maturation, (b) result in partially decreased function or total death,
and (c) affect all types of trees equally or one type selectively.
Examples are as follows.

1. *After Myofiber Maturation*

a. *One type of tree preferentially*

In myotonic atrophy (myotonia atrophica, myotonia dystrophica,
Steinert's disease), the exclusive early muscle lesion is myofiber
atrophy without significant necrosis or phagocytosis [23,35,36].
Usually it affects type I myofibers preferentially [23,35,36]. The
EMG discloses an SSAP pattern [32]. In more advanced disease, the
muscle closely resembles later neurogenic atrophy with very atrophic
fibers, pyknotic nuclear clumps, infrequent necrotic fibers, and
sometimes slight endomysial connective tissue increase [23,35]. In
myotonic atrophy the muscle enzymes are not significantly elevated
in the serum (except for occasional mild increase of creatine
phosphokinase, which is also seen in other chronic neuropathies
[23]), there is sometimes slightly or moderately elevated spinal
fluid protein [23], and there are prominent abnormalities of motoneuron endings, as shown by Coërs and Woolf [37], and McDermot [38].
All these findings are quite compatible with a chronic neurogenic
pathogenesis, which we prefer rather than a myopathic one [12,13,19,
32]. The myotonia is a series of repetitive potentials from individual myofibers. These potentials individually are indistinguishable from fibrillation potentials or positive sharp waves, both
characteristic of a denervated myofiber [39]. The myotonic potentials are generated from the abnormally labile myofiber membrane
[40] and normally the nerve controls the stability of that membrane.
We propose that a decrease of stabilizing inhibitory influence
("inhibitory factor") from the motoneuron to the myofiber allows
manifestation of myotonia [12,13,19,32]. Even the multiple internal
nuclei of myofibers seen in all stages of the disease conceivably
are compatible with a neurogenic origin, since they too occur (though
not usually so abundantly) in ordinary chronic denervation. The
ring myofibrils and sarcoplasmic masses often seen in more severely
involved muscle of myotonic atrophy [35,41-43] likewise seem
compatible with the proposed longstanding mild reduction of neural
trophic factor, allowing myofibers to undergo a form of peripheral

degeneration in which some peripheral myofibrils break and are
pushed by repeated contractions to a ring position--because such
peripheral degeneration is sometimes seen by electron microscopy in
experimentally denervated muscle fibers [23,44]. Ring myofibrils,
as well as snake-coil myofibrils, which seem to be formed in a
somewhat similar way, can be found in experimentally denervated
muscle [23,44]. Occasionally, a chronic case of typical neurogenic
atrophy has a number of myofibers with sarcoplasmic masses and
rings [23]. Perhaps the defect in myotonic atrophy more severely
involves type I motoneurons, though it could also relate to a
greater susceptibility of type I myofibers to the motoneuron abnormality. The SSAP EMG pattern of myotonic atrophy could be accounted
for by decreased flow of sap along the motoneuron tree causing
abnormality of random nerve endings (twigs) within many motor units;
such involvement of random axonal twigs would prevent activation of
a fraction of the myofibers in a motor unit and, in some patients,
cause atrophy of fibers, the activation of which would contribute
smaller spikes to the composite motor unit potential [32].

Late-onset rod disease (formerly called "rod myopathy" and
before that "nemaline myopathy") typically has mild elevation of
spinal fluid protein, occasional fasciculations and cramps, an SSAP
EMG but with additional occasional fibrillations and giant potentials, and a biopsy which in paraffin sections is usually read as
chronic denervation atrophy [23,45]. A pathogenesis involving
neural abnormality, at least partially, must therefore be considered.
In four seriously affected late-onset cases we have studied, the
degree of muscle involvement did not allow conclusive identification
of a selective myofiber type involvement. For comparison, it may
be mentioned that study of two psychotic adults who had normal
neuromuscular function clinically revealed rods in 15-20% of their
fibers located only in type II myofibers [46,47]. The central
location of this myofibrillar degeneration, i.e., rod formations,
evident in many fibers of all six adult cases is topographically

analogous to the location of central cores and targets, two other forms of myofibrillar degeneration.

b. *Both types of trees*

Diseases randomly affecting both types of neuronal trees with scattered total death after myofiber maturation seem to be the following: (a) various soma/axonal degenerations (e.g., amyotrophic lateral sclerosis [ALS]), progressive muscular atrophy, infantile spinal muscular atrophy, juvenile proximal spinal muscular atrophy, "axonal" degeneration type of peroneal muscular atrophy (Charcot-Marie-Tooth disease), and spinocerebellar degenerations, and (b) some examples of segmental denervation type of peripheral neuropathy (where the brunt of disease falls on the main trunk). These are all manifest by atrophy of both myofiber types, often with excessive oxidative enzyme staining and decreased phosphorylase in the atrophic fibers [2,25,48,49].

A partial defect involving both types of trees (motoneurons) is postulated in type II myofiber atrophy [2,12,13,25], resulting in diminution more of the trophic factor than the excitatory factor, along with a greater susceptibility of type II myofibers to that loss. Thus a partial panneuropathy is postulated to explain the type II atrophy typically seen in cachexia (with cancer or other diseases), in disuse atrophy [2,12,13,25], and in corticosteroid toxicity [50].

In myasthenia gravis we propose that there is a partial disorder in both types of motoneurons, wherein there is initially greater decrease of motoneuron excitatory factor (acetylcholine) than trophic factor, but later of both. In some instances there is sufficient loss of both to cause atrophic death of occasional myofibers [2,12,13,18]. Atrophy of myofibers has been seen in all of our cases of myasthenia gravis, sometimes preferentially of type II myofibers, but usually with some very atrophic fibers of both types that look "denervated" [49], especially in cases of longer duration [51].

Type grouping does not occur. Abnormalities of nerve endings have been shown by Coërs and Woolf [24,37] and by McDermot [52]. The proposed neuronal defect in excitatory factor could be caused by a partial disorder of the motoneuron soma resulting in decreased nourishment to random nerve twigs within each motor unit, the soma dysfunction possibly being on the basis of a mild "autoimmune" abnormality or associated therewith [18]. (Although it is also theoretically possible that in myasthenia gravis the myofiber itself initially becomes insensitive to the excitatory and trophic factors from the motoneuron, similar to curare poisoning, this seems less likely.) In juvenile myasthenia [18,32,48,53] and in typical adult myasthenia gravis [52,54,55], the SSAP EMG pattern can be explained on the basis of random junction involvement within motor units. We propose that in myasthenia gravis a qualitative or quantitative decrease of the flow of axoplasmic sap is the basis of the axonal ending (twig) abnormality and that in each motor unit it can affect some nerve endings (twigs) before others [12,13,18]. This differential susceptibility could be based on distance of an axonal twig from the sap source (soma), the angle of branching of an axonal twig, or the diameter of an axonal twig. In occasional cases of more typical flagrant motoneuron soma disease, such as ALS, one occasionally sees a "myasthenic phenomenon" on repetitive nerve stimulation and manifest by curare sensitivity [56].

It is further proposed that the facilitating myasthenic syndrome (of Lambert and Eaton) [57] may be caused by a partial defect nonselectively of motoneurons, perhaps an abnormality of the motoneuron soma. Some cases have biopsies with atrophic fibers that look somewhat denervated [23]. One case with a steroid-responsive facilitating myasthenic syndrome had her disease as a "residual" of what seems to have been a subacute motor neuropathy and whose later biopsy showed moderate type II atrophy [58]. Her EMG pattern was SSAP.

To explain the muscle weakness and atrophy of thyrotoxicosis (termed "thyrotoxic myopathy" perhaps incorrectly), a partial abnormality of all trees is proposed. The muscle shows diffuse slight

atrophy of both myofiber types, rare very atrophic fibers that look
"denervated," and no myofiber necrosis [23,59,60] (although in our
experience the myofibers are abnormally fragile and easily broken
by vigorous contraction [23]). There is no serum elevation of
muscle enzymes in thyrotoxicosis and the EMG of short small abundant
potentials [61] could be neurogenic as hypothesized above. Some
observers have even described fasciculations [62].

Because in hypokalemic periodic paralysis, atrophy of both
myofiber types, without necrosis, is seen in nearly all cases, less
in the cases of shorter duration than in ones of longer duration
[63], a slight question of at least a concomitant partial defect of
both types of motoneurons is raised. (Alternately, the atrophy
could instead represent an unresponsiveness of the myofibers to
neural trophic influences.) Type grouping does not occur [23].

In facioscapulohumeral "muscular dystrophy," where an SSAP EMG
is sometimes found in affected muscles, occasional scattered atrophic
dark angular fibers (resembling denervated fibers) are occasionally
found in minimally involved muscle, raising the slight question of
an element of neural abnormality [64]. Type grouping does not occur
[23]. This is in addition to the usual scattered, necrotic, and
regenerating fibers in minimally affected muscle which represents the
myopathic element. There appear to be occasional cases which have
primarily the pattern of neurogenic atrophy in their muscle biopsies
[23].

If myotonia does, in fact, result from decrease of a neural
inhibitory factor to the muscle (as discussed below in relation to
myotonic atrophy), then one must raise the possibility of a basically
neurogenic abnormality in myotonia congenita, perhaps partially
affecting both types of motoneurons.

2. *Before Myofiber Maturation*

Several conditions can be considered as possibly being caused
by a neural defect appearing before myofibril maturation and
affecting one type of tree (motoneuron) preferentially.

In type I hypotrophy with central nuclei [30,64], the EMG shows SSAPs. Because the basic phenomenon of congenital hypotrophy of one fiber type with central nuclei was reproduced experimentally in the rat muscle denervated at birth before myofiber maturation [30,31], it is postulated that the human disease, too, may be caused by a partial defect of virtually all type I motoneurons from before myofiber maturation. It would be only a functional neuronal defect, perhaps of a maturation factor from motoneurons, since motoneuron morphology was normal in one fatal case [30]. Type II motoneurons are likely to be normal, because a patient age 12 with this disorder has a surprising degree of useful muscle strength in spite of very small muscle bulk [64]. A neural cause is further suggested by the finding in this patient that the small type I fibers with central nuclei were all angular in cross section, some were abnormally dark with DPNH-TR, and a few had central areas histochemically resembling targets [23,64]. Theoretically, the SSAP pattern found by EMG is expected from smaller type I myofibers contributing smaller spikes to the composite motor unit potential (and perhaps occupying a smaller territory), the over-abundance of potentials resulting from a need for more units to fire in a muscle weak because of small inefficient type I myofibers. It is not known whether the type I motor units might contain a fewer than normal number of myofibers.

One hypotonic infant in our series had distinct smallness of all type I fibers without central nuclei and normal-sized type II myofibers--without evidence of central nervous system disease [23]. It is possible that his defect, in comparison with the one in type I fiber hypotrophy with central nuclei, either (a) occurred later, i.e., after the myofibers developed beyond the myotubular state with central nuclei, or (b) if earlier than that stage was milder, permitting some maturation beyond the myotubular state. This child may be a more severe form of the benign congenital hypotonia syndrome (BCH) or perhaps represents one typical aspect of BCH very early in life.

In six cases of the benign congenital hypotonia syndrome (BCH)

(a clinically nonprogressive or slowly improving disorder) studied
at ages 3-14 years, we have found some of the type I fibers to be
slightly to moderately small (probably hypotrophied) and some of
the type II fibers slightly enlarged (perhaps compensatory hypertrophy [23,65]). Possibly this change represents a partial and
mild abnormality of some of the type I motoneurons prior to myofiber
maturation, a defect that improves somewhat as the child grows.
Some of our other patients with the clinical syndrome of benign
congenital hypotonia at age 5-10 have muscle biopsies showing only
myofiber type predominance [23]. We suggest this may represent a
defective formation or very early loss of part of the population of
the opposite type of motoneuron. The cause of the proposed selective neuronal type loss in such BCH patients would seem to be nonprogressive, perhaps even monophasic before birth.

Central core disease is congenital and nonprogressive [66,67].
The patients have small muscle bulk. Even though the EMG shows
SSAP's [32], the histochemical and ultrastructural pattern of the
cores within the fibers and their limitation to type I myofibers
[65,67,68] are both strong resemblances to target fibers of ordinary
denervation [25-27]. Therefore it seems probable that central
core disease is a nonprogressive abnormality of motor innervation,
perhaps from before myofiber maturation [1]. (It could, though,
occur after myofiber maturation since in neural disorders acquired
during adulthood there can be targetoid fibers [25] which are
virtually indistinguishable from central core fibers.) In one case
there was also striking type I myofiber predominance [2,23] and in
another family there were three cases with moderate type I predominance [23,67], suggesting that at least some of the type II motoneurons were even more severely affected (i.e., lost or never
formed). To explain the pattern consisting of exuberantly abundant
short-duration normal-amplitude potentials of the EMG [32], one can
postulate a decreased number of total motor units (due to loss of
many type II's) and a somewhat smaller than normal territory of
remaining type I units (perhaps due to lack of myofiber separation

by interspersed type II myofibers). It is not known whether there
is a normal or less than normal number of myofibers per type I
motor unit. It is likely that each central core fiber has impaired
contraction efficiency and thus each unit containing core fibers
would be less efficient. To produce a given amount of work, more
of these weakened motor units would therefore need to fire and each
unit discharge more rapidly, resulting in the exuberant abundance
of potentials on slight effort.

In congenital rod (nemaline) disease, two unrelated cases had
moderate to prominent type I myofiber predominance [2,25,69,70]. The
younger girl (age 4 years) had severe type II fiber smallness (probably
II hypotrophy) without central nuclei and the other girl (age 14 years)
had only rare type II myofibers and these were hypertrophied. In
both, the rods were found only in type I myofibers. One can postulate (a) total lack of development of many type II motoneurons in
both cases (giving type I myofiber predominance), (b) poor development of the remaining type II motoneurons in the first case (giving
type II myofiber hypotrophy), and (c) in both, a defective influence
of the type I motoneurons on their myofibers which allowed the
myofibers to develop rods. The following abnormalities would contribute to the SSAP pattern found on the EMG: (a) it is postulated
that the affected myofibers and hence their motor units are less
than normally effective in contractile force; (b) some of the
myofibers are of small diameter; (c) there might be decreased
territorial extent of the type I motor units related to lack of
interspersed type II myofibers; and (d) it is not known whether
there are fewer myofibers in each motor unit. Of possible relevance
to the neurogenic hypothesis for congenital rod disease is a case
of Afiffi et al. with central core changes and rods within some
central regions [71]. Although our two cases of congenital rod
disease did not have rods preferentially located in the centers of
fibers, in six unrelated adult cases of rod disease (four late-onset
myosymptomatic [23,45], two myoassymptomatic psychotic [46,47])
there were many fibers with rods confined to a central region like
that of a central core.

VI. CONCLUSION

Although we suspect many neuromuscular diseases of erstwhile
"myopathic" or "functional" cause to be, in fact, neurogenic, a
major exception is Duchenne muscular dystrophy. In contrast to
McComas [72,73], who has recently suggested that Duchenne dystrophy
is likely to be neurogenic, we consider that scattered ischemia of
small intramuscular arterial blood vessels is the most likely pathogenesis for the circumscribed small groups of necrotic or regenerating myofibers typical of early Duchenne dystrophy [2,64,74-76].
And we have also recently proposed that another disorder, the
polymyositis/dermatomyositis complex, especially the childhood
dermatomyositis form, has a vascular pathogenesis [2], perhaps
involving small vessels of the venous circulation, and perhaps
related to immunoglobulin complexes bound to vessels .

REFERENCES

[1] Engel, W. K., "The essentiality of histo- and cytochemical
studies of skeletal muscle in the investigation of neuromuscular
disease," Neurology, 12, 778-784 (1962).

[2] Engel, W. K., "Selective and nonselective susceptibility of
muscle fiber types: A new approach to human neuromuscular disease,"
Arch. Neurol., 22, 97-117 (1970).

[3] Karpati, G. and W. K. Engel, "'Type grouping' in skeletal
muscles after experimental reinnervation," Neurology, 18, 447-455
(1968).

[4] Karpati, G. and W. K. Engel, "Transformation of the histochemical profile of skeletal muscle by 'foreign innervation,'"
Nature, 215, 1509-1510 (1967).

[5] Robbins, N., G. Karpati, and W. K. Engel, "Histochemical and
contractile properties in the cross-innervated guinea pig soleus
muscle," Arch. Neurol., 20, 318-329 (1969).

[6]   Romanul, F. C. A. and J. P. Van der Meulen, "Slow and fast muscles after cross-innervation," Arch. Neurol., 17, 387-402 (1967).

[7]   Edström, L. and E. Kugelberg, "Histochemical composition, distribution of fibres and fatiguability of single motor units," J. Neurol. Neurosurg. Psychiat., 31, 424-433 (1968).

[8]   Brandstater, M. E. and E. H. Lambert, "A histological study of the spatial arrangement of muscle fibers in single motor units within rat tibialis anterior muscle," read before the 15th annual meeting, American Association Electromyography and Electrodiagnosis, April 27, 1968, Chicago.

[9]   Doyle, A. M. and R. F. Mayer, "Studies of the motor unit in the cat," Bull. School Med. Univ. Md., 54, 11-17 (1969).

[10]   Tsairis, P., R. E. Burke, D. N. Levine, F. E. Zajac, and W. K. Engel, "Histochemical profiles of three physiologically defined types of motor units in cat gastrocnemius muscle," Neurology, 21, 436-437 (1971).

[11]   Burke, R. E., D. N. Levine, F. E. Zajac, P. Tsairis, and W. K. Engel, "Mammalian motor units: direct physiological histochemical correlation in 3 types present in cat gastrocnemius muscle," (to be published).

[12]   Engel, W. K., "Motor Neuron Disease," in The Cellular and Molecular Basis of Neurologic Disease (G. M. Shy, E. S. Goldensohn, and S. H. Appel, eds.), Lea and Febiger, Philadelphia (in press).

[13]   Engel, W. K., "Classification of Neuromuscular Disorders," in The Clinical Delineation of Birth Defects, Part VII. Muscle

(D. Bergsma, ed.), Williams and Wilkins, Baltimore, 1971.

[14]   Campa, J. and W. K. Engel, "Histochemical differentiation of motor neurons and interneurons in the anterior horn of the cat spinal cord," Nature, 225, 748-749 (1970).

[15]   Campa, J. and W. K. Engel, "Histochemistry of motor neurons and interneurons in the cat lumbar cord," Neurology, 20, 559-568 (1970).

[16] Campa, J. F. and W. K. Engel, "Histochemical and functional correlations in anterior horn neurons of the cat spinal cord," Science, 171, 198-199 (1971).

[17] Kugelberg, E. and L. Edström, "Differential histochemical effects of muscle contractions on phosphorylase and glycogen in various types of fibers: relation to fatigue," J. Neurol. Neurosurg. Psychiat., 31, 415-423 (1968).

[18] Engel, W. K. and J. R. Warmolts, "Myasthenia gravis: A new hypothesis of the pathogenesis and a new form of treatment," Ann. N.Y. Acad. Sci., (in press).

[19] Engel, W. K., "Myotonia--a different point of view," Calif. Med., 114, 32-37 (1971).

[20] Engel, W. K. and B. Horvath, "Myofibril formation in cultured skeletal muscle cells studied with antimyosin fluorescent antibody," J. Exp. Zool., 144, 209-223 (1960).

[21] Konigsberg, I. R., "Aspects of cytodifferentiation of skeletal muscle," in Organogenesis (R. L. Dehaan and H. Ursprung, eds.), Holt, Rinehart & Winston, Inc., New York, 1965, pp. 337-358.

[22] Li, C-L., W. K. Engel, and I. Klatzo, "Some properties of cultured chick skeletal muscle with particular reference to fibrillation potential," J. Cell Comp. Physiol., 53, 421-444 (1959).

[23] Engel, W. K., unpublished observations.

[24] Coërs, C. and A. L. Woolf, The Innervation of Muscle--A Biopsy Study, Blackwell Scientific Publications, Oxford, England, 1959.

[25] Engel, W. K.,"'Histochemistry of neuromuscular disease-- significance of muscle fiber types,' Neuromuscular Diseases," in Proceedings of the VIII International Congress of Neurology, Vienna, Austria, 1965, Vol. II, Excerpta Medica, Amsterdam, 1966, pp. 67-101.

[26] Engel, W. K., "Muscle target fibers--a newly recognized sign of denervation," Nature, 191, 389-390 (1961).

[27] Bajusz, E., "'Red' skeletal muscle fibers: relative independence of neural control," Science, 145, 938-939 (1964).

[28] Engel, W. K., M. H. Brooke, and P. G. Nelson, "Histochemical studies of denervated or tenotomized cat muscle: illustrating difficulties in relating experimental animal conditions to human neuromuscular diseases," Ann. N.Y. Acad. Sci., 138, 160-185 (1966).

[29] Karpati, G. and W. K. Engel, "Histochemical investigation of fiber type ratios with the myofibrillar ATPase reaction in normal and denervated skeletal muscles in guinea pig," Am. J. Anat., 122, 145-156 (1968).

[30] Engel, W. K., G. N. Gold, and G. Karpati, "Type I fiber hypotrophy and central nuclei: A rare congenital muscle abnormality with possible experimental model," Arch. Neurol., 18, 435-444 (1968).

[31] Engel, W. K. and G. Karpati, "Impaired skeletal muscle maturation following neonatal neurectomy," Develop. Biol., 17, 713-723 (1968).

[32] Warmolts, J. R. and W. K. Engel, "A critique of the 'myopathic' electromyogram," Trans. Amer. Neurol. Ass., 95, 173-177 (1968).

[33] Warmolts, J. R. and W. K. Engel, "Correlation of motor-unit behavior with histochemical myofiber type in humans by open-biopsy electromyography," Trans. Amer. Neurol. Ass.(1971) (in press).

[34] Tyler, H. R., "Physiological observations in human botulism," Arch. Neurol., 9, 661-670 (1963).

[35] Engel, W. K. and M. H. Brooke, "Histochemistry of the myotonic disorders," in Progressive Muskeldystrophie, Myotonie, Myasthenia (E. Kuhn, ed.), Springer-Verlag, Heidelberg, Germany, 1966, pp. 203-222.

[36] Brooke, M. H. and W. K. Engel, "The histographic analysis of human muscle biopsies with regard to fiber types: 3. Myotonias, myasthenia gravis and hypokalemic periodic paralysis," Neurology, 19, 469-477 (1969).

250 W. KING ENGEL AND JOHN R. WARMOLTS

[37] Coërs, C., "Les altérations du tissu musculaire et de son
innervation dans la myasthénie," in Progressive Muskeldystrophie
Myotonie Myasthenie (E. Kuhn, ed.), Springer-Verlag, New York, 1966,
pp. 325-339.

[38] McDermot, V., "The histology of the neuromuscular junction in
dystrophia myotonica," Brain, 84, 75-84 (1961).

[39] Warmolts, J. R. and W. K. Engel unpublished observations.

[40] Norris, F. H., "Intracellular recording from human striated
muscle," in Proc. 2nd Rochester Data Conf. (K. Enslein, ed.),
Pergamon Press, New York, 1963, pp. 59-72.

[41] Engel, W. K., "Chemocytology of striated annulets and sarcoplasmic masses in myotonic dystrophy," J. Histochem. Cytochem., 10,
229-230 (1962).

[42] Milhaud, M., M. Fardeau, and J. Lapresle, "Contribution à
l'étude des lésions élémentaires du muscle squelettique: ultrastructure des fibres annulaires (observées dans la dystrophie
myotonique)," Compte Rend. Soc. Biol., 158, 2274-2275 (1964).

[43] Fardeau, M., J. Lapresle, and M. Milhaud, "Contribution a
l'étude des lésions élémentaires du muscle squelettique: ultrastructure des masses sarcoplasmiques laterales (observées dans un
cas de dystrophie myotonique)," Compte Rend. Soc. Biol., 159, 15-17
(1965).

[44] Hogenhuis, L. A. H. and W. K. Engel, "Histochemistry and
cytochemistry of experimentally denervated guinea pig muscle: I.
Histochemistry," Acta Anat., 60, 39-65 (1965).

[45] Engel, W. K. and J. S. Resnick, "Late-onset rod myopathy--a
newly recognized, acquired, and progressive disease," Neurology,
16, 308-309 (1966).

[46] Engel, W. K. and H. Meltzer, "Histochemical abnormalities of
skeletal muscle in patients with acute psychoses," Science, 168,
273-276 (1970).

[47] Meltzer, H. and W. K. Engel, "Histochemical abnormalities of skeletal muscle in acutely psychotic patients," Arch. Gen. Psych. 23, 492-502 (1970).

[48] Engel, W. K., "Diseases of the neuromuscular junction and muscle," in Neurohistochemistry (C. Adams, ed.), Elsevier Publishing Co., Amsterdam, 1965, pp. 622-672.

[49] Engel, W. K. and D. E. McFarlin, "Skeletal muscle pathology in myasthenia gravis: Histochemical findings," Ann. N.Y. Acad. Sci., 135, 68-75 (1966).

[50] Pleasure, D. E., G. O. Walsh, and W. K. Engel, "Atrophy of skeletal muscle in patients with Cushing's syndrome," Arch. Neurol., 22, 118-125 (1970).

[51] Brody, I. A. and W. K. Engel, "Denervation of muscle in myasthenia gravis: Report of a patient with myasthenia gravis for 47 years and histochemical signs of denervation," Arch. Neurol., 11, 350-354 (1964).

[52] McDermot, V., "The changes in the motor end-plate in myasthenia gravis," Brain, 83, 24-39 (1960).

[53] Griggs, R. C., D. E. McFarlin, and W. K. Engel, "Severe occult juvenile myasthenia gravis responsive to longterm corticosteroid therapy," Trans. Amer. Neurol. Ass., 93, 216-218 (1968).

[54] Woolf, A. L., "Morphology of the myasthenic neuromuscular junction," Ann. N.Y. Acad. Sci., 135, 35-56 (1966).

[55] Bickerstaff, E. R. and A. L. Woolf, "The intramuscular nerve endings in myasthenia gravis." Brain, 83, 10-23 (1960).

[56] Mulder, D. W., E. H. Lambert, and L. M. Eaton, "Myasthenic syndrome in patients with amyotrophic lateral sclerosis," Neurology, 9, 627-631 (1959).

[57] Lambert, E. H., L. M. Eaton, and E. D. Rooke, "Defect of neuromuscular conduction associated with malignant neoplasms," Amer. J. Physiol., 187, 612-613 (1956).

[58] Vroom, F. Q. and W. K. Engel, "Nonneoplastic steroid responsive Lambert-Eaton myasthenic syndrome," Neurology, 19, 281 (1969).

[59] Engel, W. K. and Brooke, M. H., "Muscle biopsy in ALS and other motor neuron diseases," in Motor Neuron Diseases (F. H. Norris Jr. and L. T. Kurland, eds.), Grune and Stratton, New York, 1969, pp. 154-159.

[60] Engel, W. K., "Muscle Biopsy," in a symposium "Review of Current Concepts of Myopathies," Clinical Orthopaedics and Related Research (W. K. Engel, guest ed.), Vol. 39, Chap. 7, Philadelphia, 1965, pp. 80-105.

[61] Harvard, C. W. H., E. D. R. Campbell, H. B. Ross, and A. W. Spence, "Electromyographic and histological findings in the muscle of patients with thyrotoxicosis," Quart. J. Med., 32, 145-163 (1963).

[62] McEachern, D. and W. E. Ross, "Chronic thyrotoxic myopathy," Brain, 65, 181-192 (1942).

[63] Griggs, R. C., W. K. Engel, and J. S. Resnick, "Acetazolamide treatment of hypokalemic periodic paralysis," Ann. Intern. Med., 73, 39-48 (1970).

[64] Engel, W. K., "Muscle biopsies in neuromuscular diseases," Pediat. Clin. N. Amer., 14, 963-995 (1967).

[65] Engel, W. K., "A critique of congenital myopathies and other disorders," in Exploratory Concepts in Muscular Dystrophy and Related Disorders (A. T. Milhorat, ed.), Excerpta Medica Foundation, New York, 1967, pp. 27-40.

[66] Shy, G. M. and K. R. Magee, "A new congenital non-progressive myopathy," Brain, 79, 610-621 (1956).

[67] Engel, W. K., J. B. Foster, B. P. Hughes, H. E. Huxley, and R. Mahler, "Central core disease--an investigation of a rare muscle cell abnormality," Brain, 84, 167-185 (1961).

[68] Resnick, J. S. and W. K. Engel, "Target fibers--structural and cytochemical characteristics and their relationship to neurogenic

muscle disease and fiber types," in <u>Exploratory Concepts in Muscular Dystrophy and Related Disorders</u> (A. T. Milhorat, ed.), Excerpta Medica Foundation, New York, 1967, pp. 255-266.

[69] Shy, G. M., W. K. Engel, J. E. Somers, and T. Wanko, Nemaline myopathy--a new congenital myopathy," <u>Brain</u>, 86, 793-810 (1963).

[70] Engel, W. K., T. Wanko, and G. M. Fenichel, "Nemaline myopathy, a second case," <u>Arch. Neurol.</u>, 11,22-39 (1964).

[71] Afifi, A. K., J. W. Smith, and H. Zellweger, "Congenital nonprogressive myopathy: central core disease and nemaline myopathy in one family," <u>Neurology</u>, 15, 371-381 (1965).

[72] McComas, A. J. and R. E. P. Sica, "Muscular dystrophy: myopathy or neuropathy," <u>Lancet</u>, i, 119 (1970).

[73] McComas, A. J., R. E. P. Sica, and S. Currie, "Muscular dystrophy: evidence for a neural factor," <u>Nature</u>, 226, 1263-1264 (1970).

[74] Hathaway, P. W., W. K. Engel, and H. Zellweger, "Experimental myopathy after microarterial embolization," <u>Arch. Neurol.</u>, 22, 365-378 (1970).

[75] Mendell, J. R., W. K. Engel, and C. Derrer, "Duchenne muscular dystrophy: Functional ischemia reproduces its characteristic lesions," <u>Science</u>, (in press).

[76] Fardeau, M., "Etude ultrastructurale des differents types de fibres musculaires dans la musculature squelettique du cobaye," <u>Proc. Conf. Marseille, Oct. 28-Nov. 1, 1970</u> (in press).

CHAPTER 9

STRUCTURAL CHANGES IN HUMAN AND CHICKEN MUSCULAR DYSTROPHY

S. Ahmad Shafiq, Valerie Askanas,
Stephen A. Asiedu, and Ade T. Milhorat

Institute for Muscle Disease
New York, New York

I.   INTRODUCTION

The Duchenne type of muscular dystrophy is characterized by progressive wasting and weakness of muscles with onset in early childhood and sex-linked inheritance. The etiology of the disease is not known. Though generally considered to be a primary myopathy, suggestions have been made that the primary lesion may, in fact, be in the microcirculation [17], in the nervous system [21], or in the connective tissue [6]. The hereditary muscular dystrophy in chicken [4] is an autosommal disease but many histopathological similarities with the human dystrophy have been noted [20]. The purpose of this report is to discuss the histopathological features of muscles from these two dystrophies with special reference to early changes.

II.   DUCHENNE MUSCULAR DYSTROPHY

The gross degenerative changes in muscles of patients with

Duchenne type of muscular dystrophy are well known since the studies
of Erb [13]; several fine structural studies have also been published recently [18,22,23]. The typical histopathological picture
consists of atrophy and splitting of muscle fibers with a marked
variation of fiber diameters and infiltration of affected muscles
with fat and connective tissue [Figs. 1(a) and (b)]. In young
patients, before the general architecture of the muscle has been
grossly disrupted, necrosis of fibers scattered singly or in small
groups is also a common feature [Fig. 1(c)]. Such fibers usually
have hyaline fiber contents and are infiltrated with phagocytes
[Fig. 1(d)]. By electron microscopy a fine granular material with
occasional remnants of cytoplasmic organelles is seen in them; the
plasma membranes of such fibers are also generally disintegrated so
that only the basement membranes are left on the surface [Fig. 2(a)].
The phagocytic cells inside the necrotic fibers have rough-surfaced
endoplasmic reticulum, prominant lysosomal structures, and usually
their plasma membranes are intact [Fig. 2(a)]. In some fibers,
however, the plasma membranes of the phagocytic cells appear broken
and their lysosomes and other contents lie freely in the hyaline
fibers [Fig. 2(b)]. In many other fibers without infiltrated
phagocytes, lysosomes are, none the less, found especially around
the fiber nuclei [Fig. 2(c)].

In relation to these degenerative changes two questions, namely,
the origin of phagocytic cells and source of lysosomes in the
dystrophic fibers are of interest. In muscles experimentally
injured by heat or cold shock, Allbrook [2] and Price et al. [27]
proposed that the myoblasts which appear in the necrotic fibers
probably act as phagocytes at first; after the necrotic debris has
been removed they begin their myogenic activity. In dystrophic
patients, Adams et al. [1] maintained that active phagocytosis of
degenerating fibers was extremely rare, but several workers [14,19,
22,26] have now found it commonly in the early stages of the disease.
In the absence of a prominant inflammatory response in muscles of
patients with Duchenne type of muscular dystrophy, Gilbert and

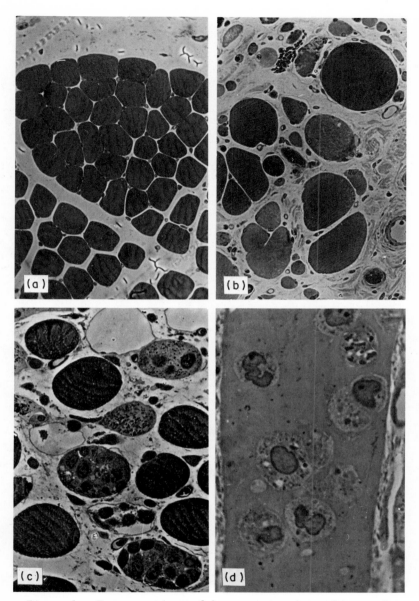

Figure 1. Epoxy sections of human vastus lateralis muscle stained with methylene blue. (a) Eight-year-old normal boy; the fibers have a polygonal outline (190 x). (b) Eight-year-old patient of Duchenne dystrophy showing marked variation of fiber diameters (190 x). (c) Three-year-old patient of muscular dystrophy; several necrotic fibers contain phagocytes (480 x). (d) Six-year-old dystrophic boy; high magnification of a hyaline fiber showing cellular infiltration (1200 x).

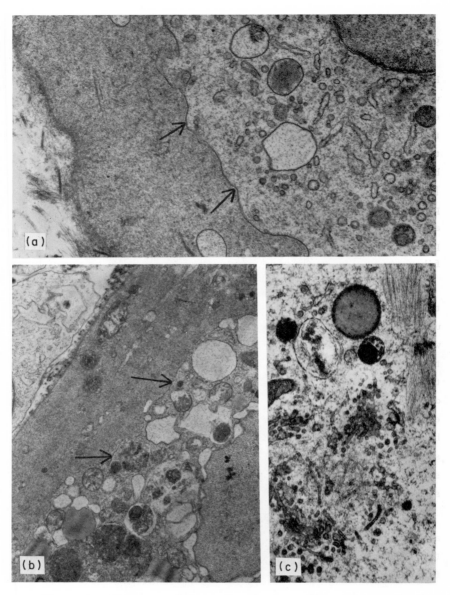

Figure 2. Electron micrographs of vastus lateralis muscle from dystrophic patients. (a) Hyaline fiber with part of a phagocyte (right side of the picture) in which the plasma membrane is intact (indicated by arrows) (17,500 x). (b) Another hyaline fiber with phagocyte whose plasma membrane (indicated by arrows) is broken (10,500 x). (c) Lysosomes and enlarged Golgi field in a fiber in which infiltration by phagocytes was not seen (25,000 x).

Hazard [15] proposed that the phagocytes seen in the necrotic fibers
were endogenous myoblasts.   In our study, however, we have commonly
encountered single cells lying in the interstitial spaces between
the dystrophic muscle fibers [Fig. 1(c)].   By electron microscopy
these cells show prominent endoplasmic reticulum, lysosomes, and
bundles of thin filaments coursing irregularly in the cytoplasm
[Fig. 3(a)].   Morphologically they appear to be histiocytes and
probably they infiltrate the degenerating fibers for phagocytic
activity.   Infiltration of cells is clearly demonstrated by the
presence of occasional polymorphonuclear leucocytes inside necrotic
fibers [Fig. 3(b)].

    The origin of lysosomal enzymes in dystrophic muscles is con-
troversial.   Tappel et al. [29] in dystrophic animal muscles attributed the increase of enzyme activity to invading macrophages,
whereas, Weinstock and Iodice [31] consider it as originating in
the lysosomes of the muscle cells themselves.   As mentioned above
the phagocytic cells are common in early stages of Duchenne
dystrophy and probably represent an important source of lysosomal
enzymes.   In late stages of the disease, however, when phagocytosis
has subsided, lysosomes inside the muscle fibers are still commonly
seen [22]; it would therefore appear that, at least in the late
stages, lysosomes of muscle fibers themselves are probably more
significant than those of invading cells.

    The early changes in muscular dystrophy seem to be related to
alterations in histochemical and physiological properties of fiber
types.   Dubowitz [11] and Bell and Conen [5] have described muscle
fibers from biopsy specimens of dystrophic patients which reacted
strongly with both oxidative and glycolytic enzymes.   Prolongation
of contraction times correlated with increased proportion of fibers
rich in oxidative enzymes has also been noted [8].   These changes
have generally been considered to represent defective differentiation of fiber types [8,11].   For identification of fiber types in
human muscles Engel [12] has favored the myofibrillar adenosine
triphophatase (ATPase) reaction [25] whose staining pattern is well

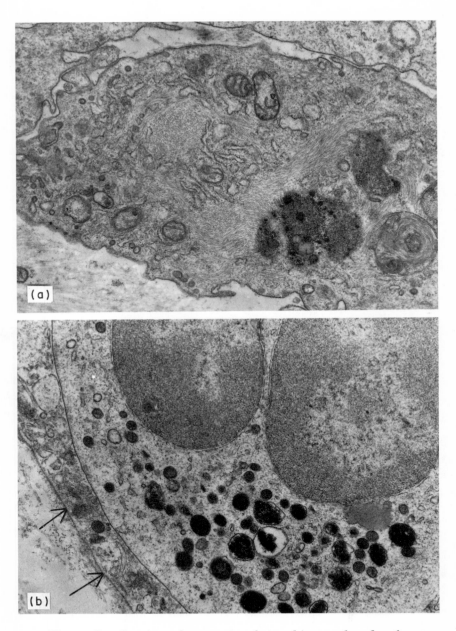

Figure 3. Sections from vastus lateralis muscle of a 6-year-old dystrophic patient (17,500 x). (a) Micrograph of a histiocyte from interstitial tissue of the muscle. (b) Part of a polymorpho-nuclear leucocyte inside a necrotic muscle fiber; the basement membrane of the fiber is indicated by arrows.

retained in many pathological conditions in comparison to those for
oxidative and glycolytic enzymes. More recently Guth and Samaha
[16] and Brooke and Kaiser [7] have used the ATPase technique in
combination with acid (pH 4.35) and base (pH 10.4) preincubations
for classification of fiber types. Figs. 4(a)-(d) illustrate changes
in these reactions in the biopsy specimen from a patient with chronic
neurogenic condition following poliomyelitis. It can be readily seen
by phosphorylase [Fig. 4(a)] and succinic dehydrogenase [Fig. 4(b)]
tests that all fibers are reacting comparatively uniformly and fiber
types are not clearly distinguishable. In serial sections from the
same specimen, the reactions for myofibrillar ATPase clearly distinguish the fiber types; type II fibers reacting more strongly than
type I following preincubation at pH 10.4 [Fig. 4(c)], the reverse
being true after preincubation at pH 4.35 [Fig. 4(d)]. Figures 4(e)
and (f) illustrate these ATPase reactions from the biopsy specimen of
a 7-year-old patient with Duchenne muscular dystrophy. The muscle
shows marked variation of fiber diameters but as in the previous
case the ATPase reactions with acid and base preincubations distinguish the two fiber types with typical reciprocal staining
characteristics. Thus, it appears that though reactions with
respect to oxidative and glycolytic enzymes are abnormal in Duchenne
dystrophy [5,11], the fiber type differentiation with respect to
ATPase reactions is not grossly altered.

III. CHICKEN MUSCULAR DYSTROPHY

The histopathology in chicken dystrophy has been examined in
detail by Julian and Asmundson [20]. The changes in affected
muscles include increased variation of fiber diameters, increase in
number of fiber nuclei, vacuolation, fiber necrosis with phagocytosis
and fat infiltration. From electron microscopic observations,
thinning of fibrils due to loss of the myofilaments appears to be a
common reaction. In the normal muscle, the fibrils are closely
packed [Fig. 5(a)], but in many dystrophic fibers the fibrils,
though retaining their general striation pattern, become widely

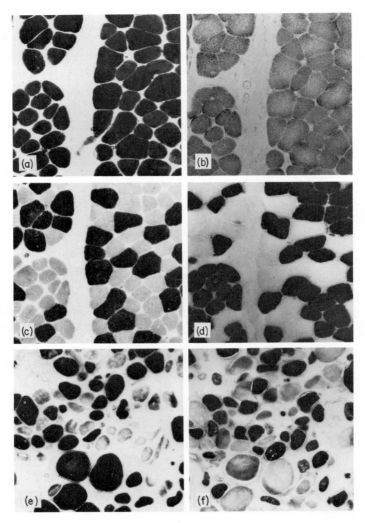

Figure 4. Histochemical reactions of deltoid muscle from a patient with chronic neurogenic condition after poliomyelitis (a)-(d) and of vastus lateralis muscle from a patient with Duchenne dystrophy (e), (f) (120 x). (a) Phosphorylase. (b) Succinic dehydrogenase. (c) Myofibrillar ATPase following pH 10.4 preincubation. (d) ATPase, pH 4.35 preincubation. (e) ATPase, pH 10.4 preincubation. (f) ATPase, pH 4.35 preincubation.

separated, are thinner, and show frayed myofilaments on their
surface [Fig. 5(b)]. In many fibers simple lysosomes [Fig. 6(a)] or
vacuoles with membraneous contents [Fig. 6(b)] are also present.
Swollen mitochondria as well as vacuoles which probably represent

Figure 5. Electron micrographs of pectoralis major muscle of
5-month-old chicken (22,500 x). (a) Normal. (b) Dystrophic.

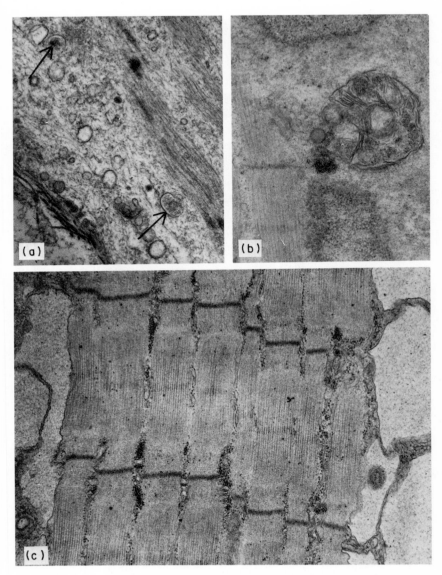

Figure 6.   Electron micrographs of pectoralis major from a
1-year-old dystrophic chicken (22,500 x).   (a) Fiber showing simple
lysosomes (indicated by arrows).   (b) Fiber with a vacuole containing membraneous material.   (c) Fiber with distended elements of
sarcoplasmic reticulum.

distended portions of sarcoplasmic reticulum [Fig. 6(c)] are also a
common feature.   These changes, however, do not seem to be specific
for the dystrophic lesion and have been observed in a number of
unrelated muscle diseases in patients [18].

The question of fiber types in chicken dystrophy has been
examined by several workers.   It has been shown that white muscles,
e.g., pectoralis major, composed of type II fibers are more susceptible to the dystrophic process than the red muscles [9,20,28,32].
Moreover, the detailed studies of Cosmos [9,10] have demonstrated
defective maturation of white fiber characteristics in respect to
the activities of phosphorylase and succinic dehydrogenase enzymes.
However, as far as the myofibrillar ATPase reactions are concerned,
Figs. 7(a)-(d) illustrate that, as in human dystrophy, they are not
materially altered.   The fibers in the normal pectoralis muscle,
being almost entirely of type II, give a strong positive reaction
following preincubation at pH 10.4 [Fig. 7(a)] and a negative
reaction after incubation at pH 4.35 [Fig. 7(b)]; in the dystrophic
pectoralis the fibers seem to react in a similar pattern [Fig. 7(c)
and (d)].   Also in respect to the fine structural characteristics
of fiber types the changes are not striking.   In the normal chicken,
type I and type II fibers appear to differ mainly in the thickness
of Z lines, the sarcoplasmic reticulum being comparatively sparse
in both fiber types [28].   In the dystrophic pectoralis muscle the
Z band thickness in many fibers is 500-550 Å which appears normal
for the type II fibers; in some dystrophic fibers, however, it is
about 700 Å [Fig. 8(a)], which is comparable to that in immature
pectoralis fibers of the newly hatched chicken [Fig. 8(b)] and is
appreciably less than that in type I fibers [Fig. 8(c)].

The problem of the earliest histological changes was previously
studied in the sartorius of chicken which was found to be mildly
involved by the dystrophic process.   The earliest difference between
the normal and dystrophic muscles appeared to be an increased number
of fibers in the latter [28].   Currently, we are doing fiber counts

from the pectoralis muscles of normal and dystrophic embryos, and preliminary results indicate that, as in the sartorius, there is a greater number of muscle fibers in the dystrophic pectoralis (Fig. 9).

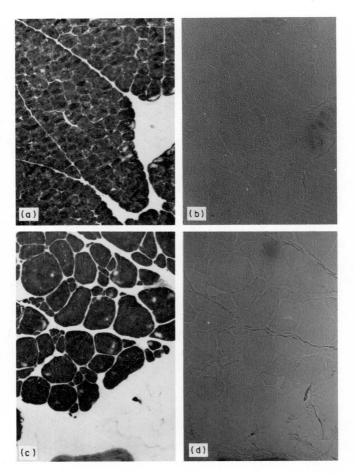

Figure 7. Histochemical reactions of the pectoralis major of 5-month-old chicken (120 x). (a) Normal muscle, myofibrillar ATPase reaction after pH 10.4 preincubation. (b) Normal muscle, ATPase reaction, pH 4.35 preincubation. (c) Dystrophic muscle, ATPase reaction, pH 10.4 preincubation. (d) Dystrophic muscle, ATPase reaction, pH 4.35 preincubation.

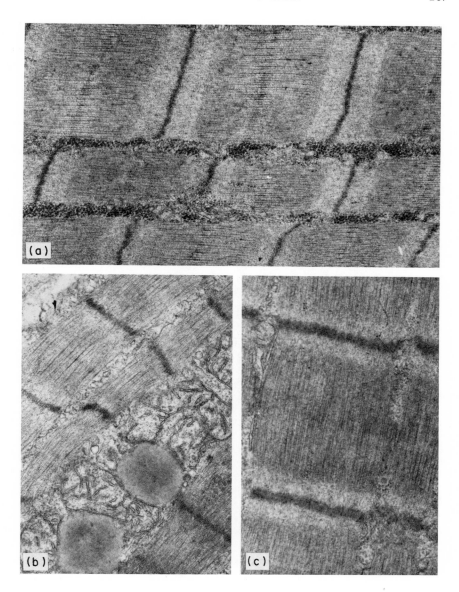

Figure 8. Electron micrographs of longitudinal sections of chicken muscles (22,500 x). (a) 1-year-old dystrophic chicken, pectoralis major. (b) Newly hatched normal chicken, pectoralis major. (c) 1-year-old normal chicken, type I fiber of sartorius muscle.

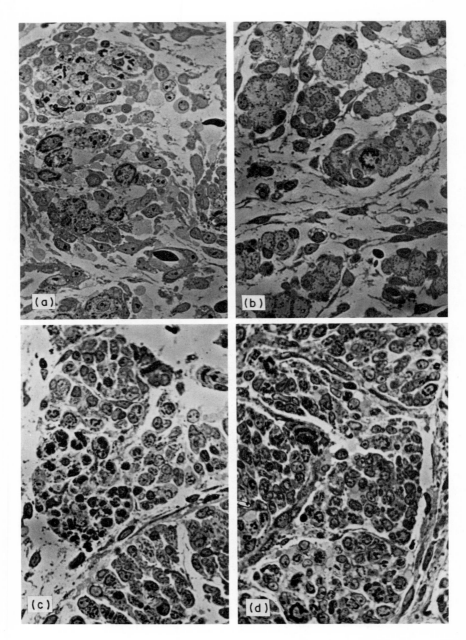

Figure 9. Transverse sections of pectoralis muscles from
chicken embryos (520 x). (a) Ten-day-old normal embryo. (b) Ten
day dystrophic embryo. (c) Fifteen day normal embryo. (d) Fifteen
day dystrophic embryo; fibers in muscles of dystrophic embryos
appear to be more numerous.

These results support the findings of greater production of muscle
fibers in dystrophic cultures [3] as well as of increased DNA content in muscles of dystrophic embryos [24,30]; however, the significance of this feature to primary lesion in muscular dystrophy is
not clear at present.

IV. SUMMARY

The histological and fine structural changes related to
degeneration of fibers in human (Duchenne) and chicken muscular
dystrophy are described. The histochemical reactions of fiber types
of dystrophic muscles are also illustrated; the results show that
unlike the abnormal reactions for oxidative and glycolytic enzymes
reported previously, the myofibrillar ATPase reactions (following
acid and base preincubations) are not significantly altered. It is
proposed that the earliest histological change judged by comparison
of normal and dystrophic chicken embryos is in the presence of
increased number of fibers in dystrophic muscles.

REFERENCES

[1]   Adams, R. D., D. Denny-Brown, and C. M. Pearson, Diseases of
Muscle, Harper & Brothers, New York, 1962, p. 735.

[2]   Allbrook, D. B., "An electron microscopic study of regenerating
skeletal muscle," J. Anat. (London), 96, 137-152 (1962).

[3]   Askanas, V., S. A. Shafiq, and A. T. Milhorat, "Normal and
dystrophic chicken muscle at successive stages in tissue culture,"
Arch. Neurol. (Chicago), 24, 259-265 (1971).

[4]   Asmundson, V. S., and L. M. Julian, "Inherited muscle abnormality in the domestic fowl," J. Hered., 47, 248-252 (1956).

[5]   Bell, C. D., and P. E. Conen, "Histochemical fiber 'types' in
Duchenne muscular dystrophy," J. Neurol. Sci., 10, 163-171 (1970).

[6]   Bourne, G. H., and M. N. Golarz, "Human muscular dystrophy as

an aberration of the connective tissue," Nature (London), 183, 1741-1743 (1959).

[7]   Brooke, M. H., and K. K. Kaiser, "Muscle fiber types: How many and what kind?" Arch. Neurol. (Chicago), 23, 369-379 (1970).

[8]   Buchthal, F., H. Schmalbruch, and Z. Kamieniecka, "Contraction times and fiber types in patients with progressive muscular dystrophy," Neurology (Minneap.), 21, 131-139 (1971).

[9]   Cosmos, E., "Enzymatic activity of differentiating muscle fibers. I. Development of phosphorylase in muscles of domestic fowl," Develop. Biol., 13, 163-179 (1966).

[10]   Cosmos, E., and J. Butler, "Differentiation of fiber types in muscles of normal and dystrophic chickens. A quantitative and histochemical study of the ontogeny of muscle enzymes," in Exploratory Concepts in Muscular Dystrophy and Related Disorders, (A. T. Milhorat, ed.), Excerpta Medica Foundation, Amsterdam, 1967, p. 197-204.

[11]   Dubowitz, V., Developing and Diseased Muscle, Spasties International Medical Publication in association with Heinemann Medical Books, London, 1968, p. 107.

[12]   Engel, W. K., "The essentiality of histo- and cytochemical studies of skeletal muscle in the investigation of neuromuscular disease," Neurology (Minneap.), 12, 778-794 (1962).

[13]   Erb, W., "Dystrophia muscularis progressiva: Klinische und pathologisch - anatomische studien," Deut. Z. Nervenheilk., 1, 13-94, 173-261 (1891).

[14]   Gilbert, R. K., and W. A. Hawk, "The incidence of necrosis of muscle fibers in Duchenne type muscular dystrophy," Am. J. Pathol., 43, 107-122 (1963).

[15]   Gilbert, R. K., and J. B. Hazard, "Regeneration in human skeletal muscle," J. Path. Bacteriol., 89, 503-512 (1965).

[16]   Guth, L., and F. J. Samaha, "Qualitative differences between

actomyosin ATPase of slow and fast mammalian muscle," Exptl. Neurol., 25, 138-152 (1969).

[17] Hathaway, P. W., W. K. Engel, and H. Zellweger, "Experimental myopathy after micro arterial embolization," Arch. Neurol. (Chicago), 22, 365-378 (1970).

[18] Hudgson, P., and G. W. Pearce, "Ultramicroscopic studies of diseased muscle," in Disorders of Voluntary Muscle (J. N. Walton, ed.), Little, Brown and Company, Boston, 1969, p. 277-317.

[19] Hudgson, P., G. W. Pearce, and J. N. Walton, "Preclinical muscular dystrophy: histopathological changes observed on muscle biopsy," Brain, 90, 565-576 (1967).

[20] Julian, L. M., and V. S. Asmundson, "Muscular dystrophy of the chicken," in Muscular Dystrophy in Man and Animals, (G. H. Bourne and M. N. Golarz, eds.), Karger, Basel, 1963, p. 457-498.

[21] McComas, A. J., R. E. P. Sica, and S. Currie, "Muscular dystrophy: Evidence for a neural factor," Nature (London), 226, 1263-1264 (1970).

[22] Milhorat, A. T., S. A. Shafiq, and L. Goldstone, "Changes in muscle structure in dystrophic patients, carriers and normal siblings seen by electron microscopy; correlation with levels of serum creatinephosphokinase (CPK)," Ann. N.Y. Acad. Sci., 138, 246-292 (1966).

[23] Mölbert, E., "Feinstrukturelle Veränderungen bei der Muskel dystrophie," in Die progressiv-dystrophischen Myopathien, (H. Heyck and G. Laudahn, ed.), Springer-Verlag, Berlin, 1969, p. 119-138.

[24] Morgan, D. F., and H. Herrmann, "Comparison of muscle tissue from normal and dystrophic chick at different stages of development," Proc. Soc. Exp. Biol. Med., 120, 68-72 (1965).

[25] Padykula, H. A., and E. Herman, "The specificity of the histo-chemical method for adenosine triphophatase," J. Histochem. Cytochem., 3, 170-195 (1955).

[26] Pearce, G. W., and J. N. Walton, "Progressive muscular dystrophy: the histopathological changes in skeletal muscle obtained by biopsy," J. Pathol. Bacteriol., 83, 535-550 (1962).

[27] Price, H. M., E. L. Howes, and J. M. Blumberg, "Ultrastructural alterations in skeletal muscle injured by cold. II. Cells of the sarcolemma tube: observations on 'discontinuous' regeneration and myofibril formation," Lab. Invest., 13, 1279-1302 (1964).

[28] Shafiq, S. A., V. Askanas, and A. T. Milhorat, "Fiber types and preclinical changes in chicken muscular dystrophy," Arch. Neurol. (Chicago), 25, 560-571 (1971).

[29] Tappel, A. L., H. Zalkin, K. A. Caldwell, I. D. Desai, and S. Shibko, "Increased lysosomal enzymes in genetic muscular dystrophy," Arch. Biochem. Biophys., 96, 340-346 (1962).

[30] Weinstock, I. M., "Comparative biochemistry of myopathies," Ann. N.Y. Acad. Sci., 138, 199-212 (1966).

[31] Weinstock, I. M., and A. A. Iodice, "Acid hydrolase activity in muscular dystrophy and denervation atrophy," in Lysosomes in Biology and Pathology (J. T. Dingle and H. B. Fell, eds.), North-Holland, Amsterdam, 1969, Vol. I, p. 450-468.

[32] Wilson, B. W., M. A. Kaplan, W. C. Merhoff, and S. S. Mori, "Innervation and the regulation of acetyl cholinesterase activity during the development of normal and dystrophic chick muscle," J. Exptl. Zool., 174, 39-54 (1970).

CHAPTER 10

OXYGEN UPTAKE BY STRIATED MUSCLE

W. N. Stainsby and J. K. Barclay

University of Florida
Department of Physiology
College of Medicine
Gainesville, Florida

I. INTRODUCTION

Current understanding for most devotees of muscle energetics is based on an immense literature of myothermic and chemical studies mainly on amphibian muscle. This has been reviewed recently by Mommaerts [16]. The small quantity of literature which exists to date that estimates muscle energetics from measurements of oxygen uptake is usually omitted from such reviews. We believe there are now a sufficient number and variety of studies on oxygen uptake to establish the energetic patterns derived from such studies. It is the intent of this short review to examine these patterns, especially the differences, and to ask about their meaning. It will be amply clear to the reader that we, at least, have few answers. Hopefully more future work will be directed to this end.

II. FROG SKELETAL MUSCLE

The first measure of net oxygen uptake for contractions, above resting values, was by Fenn [9]. The mechanics of these tetanic contractions were not measured or altered. Three observations in the paper seem of special interest. (a) Recovery oxygen did not increase linearly with increasing stimulus durations, there being relatively less used for the longer durations. (b) While resting oxygen uptake showed a large change with temperature, the oxygen uptake for the contractions did not show a measurable temperature effect. (c) Oxygen uptake per stimulus at 12-22°C was 0.1 to 0.2 µl/g wet weight per stimulus if the Harvard inductorium was providing stimuli at a frequency near 50 cps.

The first report of the oxygen uptake for isotonic and isometric twitch contractions is that by Fischer [10]. The contractions were afterloaded, presumably with fixed initial or rest length before each contraction series. Load was increased in steps until the contractions were isometric. Net oxygen uptake increased with load but at a decreasing rate, being maximal at a high load which was less than the isometric load. The isometric load was associated with a less than maximal oxygen uptake, but was, however, higher than that for the smallest load of the series. These experiments were done at about 12°C and were interpreted in the light of the evidence at the time, based mainly on myothermic experiments, that the energy exchange was dependent upon two factors, (a) the length of the fibers during contraction and (b) the mechanical work done. If our calculations are correct the oxygen uptake was about 0.23 µl/g twitch.

During the next 25 years, while myothermic and chemical studies at 0°C advanced in precision and reproductability, there seem to have been no studies clearly relating oxygen uptake to mechanical variables during contraction.

The next measurements of oxygen consumption by frog muscle at 27.5°C was by Whalen and Collins [24] who reported that oxygen uptake increased with load but did not decrease when the contractions became isometric. The observation that the oxygen uptake for the isometric contraction did not decrease resulted in mechanical work not having a significant relationship with oxygen uptake. The authors reported the difference between their studies and those of Fischer [10] to be due, most likely, to the difference in the temperature used in the experiments. This suggestion was based on an older report from myothermic experiments by Fischer [11], where it was shown that the now classical heat relationships were distorted and then lost when the heat measurements were made at progressively higher temperatures. In view of their results and in accord with other observations at the time, Whalen and Collins [24] concluded that net oxygen uptake for contraction was determined mainly by (a) rest length and (b) by load or tension on the muscle. Oxygen uptake for contractions was in the range of 0.2 to 0.3 µl/g twitch.

Baskin and Gaffin [3] reported oxygen uptake for isometric contractions of frog sartorius at 10°C to be related to tension. The highest oxygen uptake and tension were observed near rest length; optimum length and rest length appeared to be the same. It was concluded that tension determined the oxygen uptake above a minimal level referred to as activation energy. No effort was made to decide between contraction length and tension as the cause of oxygen uptake changes. In contrast to Fischer [10] it appeared to be assumed that length per se had no effect.

Baskin [1] reported oxygen uptake to be related to load and shortening in isotonic contractions. In these experiments it appears that rest length was always near L_o, but shortening was limited at each load by a stop. Presumably the tension rose above the lever load when the lever hit the stop and the contraction became isometric. Although some simple isometric contractions were studied, no accounting of the tension development during the isometric part of these partially isotonic contractions seems to have been made. We wonder if this isometric tension development has accounted for some of the oxygen uptake for the contractions.

In another report Baskin reports a weak relationship between oxygen uptake and velocity of contraction [2]. He also presents data suggesting external work and oxygen uptake are poorly related. However, the contraction mechanics of these studies were particularly complex and,as mentioned above, interpretation of the data is very difficult.

In these three papers the oxygen uptake for contraction was in the range of 0.5 to 2.0 µl/g twitch which is much higher than the preceding reports. We have no certain explanation for the difference. One possible explanation might be related to the size of the stimulus used, 10 V of 5 msec duration. We wonder if such long and powerful stimulation could have caused some repetitive contractions in some of the muscle fibers.

From the studies mentioned in this section the factors which seem to be related to oxygen uptake and hence energy exchange are tension development, length during contraction, and shortening.

III. MAMMALIAN HEART MUSCLE

The early literature about oxygen consumption of the heart is extremely complex, and the conclusions reached were variable. The end of this period was probably heralded by the report of Sarnoff et al. [18]. Their findings have stood repeating by numerous other groups. Put as simply as possible, they observed that myocardial oxygen uptake per beat increased when blood pressure was increased, but oxygen uptake increased little when stroke volume was increased, especially if peak systolic pressure was kept constant. Reasonable fits for common denominators appeared to be tension-time integral per beat or peak systolic pressure. It is risky to translate heart function into the usual muscle terms, but if resting heart volume is relatively constant then stroke volume becomes proportional to shortening and pressure to tension or load. In these terms load appears to be the major determinant of oxygen uptake per contraction of the heart and rest length and shortening appear to be of little importance.

More detailed analyses of oxygen uptake by heart muscle for contractions has involved use of papillary muscles or strips of myocardium. The methodology seems to have been established about 1960 and is exemplified in a paper by Lee [13]. It is difficult to dissect out the usual mechanical variables in this paper, but it appears to us that the extra oxygen uptake per contraction was largely determined by resting tension. This oxygen uptake was about 0.17 μl/g twitch at 37°C.

A paper by Whalen in 1961 [23] is presented in more conventional terms. Oxygen uptake was measured at 31.7°C under a variety of mechanical contraction conditions. The basic relationship for afterloaded contractions between oxygen uptake and load is similar to that of Fischer [10] for frog sartorius. However, in the light of other relationships presented, it was concluded that the major determinant of contraction oxygen uptake was rest (initial) length or mean length during each contraction. Tension, work, and shortening were considered to be unrelated to energy exchange. Oxygen uptake was about 0.6 μl/g twitch.

McDonald [15] found that contractile element work (internal plus external work) was well correlated with oxygen uptake per contraction. Developed tension changes also appeared to correlate well with oxygen uptake. Only one type of contraction was studied. Therefore no attempt was made to test the possibility that these correlations were fortuitous and not indications of cause and effect. Oxygen uptake at 29°C was in the range of 1.0 μl/g per contraction.

Coleman [5] using similar preparations and methods came to similar conclusions regarding contractile element work and oxygen uptake. The correlation seemed good only in the middle load range. The experiments shown in his Fig. 3 are not shown in Fig. 4. We wonder how the fit would be. Other aspects of the results are similar to those of McDonald [15]. Oxygen uptake per beat was 0.3-2.5 μl/g beat.

A report by Coleman et al. [6], again using similar preparations and methods, also generally agrees with the contractile element work hypothesis. They added another experimental procedure. They measured the oxygen uptake for isometric contractions with initial length reduced sufficiently to produce the same tension as their isotonic contractions. For a given tension the isometric contractions required less oxygen than the isotonic contractions which shortened and performed external work. Since load or force were the same, internal work was the same. They ascribed the extra oxygen uptake to the external work done. For some reason they also calculated internal work in their experiments and plotted the contractile element "internal" work and the contractile element "external" work. The slope of each against the oxygen uptake ascribed to them are significantly different.

These authors do not question the results or pose any other possibilities. For example, shortening and initial length have both been suggested at one time or another to directly affect oxygen uptake. Shortening was much reduced because the contractions were isometric. Initial length was reduced to create tension equal to the isotonic contractions. The reduction in oxygen uptake could have been due to the change in shortening or the decreased initial length or to both factors together.

IV. MAMMALIAN SKELETAL MUSCLE

Mammalian skeletal muscle has been studied very little in regard to oxygen uptake and contraction mechanics. There have been numerous studies of oxygen uptake during and after contractions. These have been well summarized by Fales et al. [8]. We suppose the lack of these muscles to be due to the technical difficulties. It is necessary to measure blood flow and blood oxygen concentrations; existing lever systems were totally inadequate for such large and rapidly contracting muscles.

Using a pneumatic lever, Fales et al. [8] studied oxygen uptake for brief isotonic tetanic contractions. Unfortunately, stimulus frequency, duration of stimulus trains (the number of impulses per

contraction), load, and rest length were all varied in an almost
random manner. Oxygen uptake was not related to work, shortening,
and load. Oxygen uptake appeared to be dependent mainly on the
number of impulses delivered to the muscle, and this relationship
appeared to be altered if the stimulus frequency in each train was
altered. It can also be noted in the data that the oxygen uptake
for a given number of impulses is lower if they are delivered continuously than when delivered in short trains with a brief rest
period between trains. Oxygen uptake per stimulus was about 0.5
$\mu l/g$.

DiPrampero et al. [7] measured oxygen uptake for tetanic contraction but the mechanics of the contractions were not measured.
The results cannot be compared.

Using the same muscle preparation and a copy of the pneumatic
muscle lever, Stainsby [21] measured the oxygen uptake for isotonic
and isometric twitch contractions. For these single twitches oxygen
uptake was around 0.5 $\mu l/g$ twitch. In the gastrocnemius-plantaris
muscle group of the dog, the relationship between oxygen uptake and
load, or tension, was the same for isometric as for freeloaded and
afterloaded isotonic contractions. For a given load or tension,
oxygen uptake was the same whether the muscle shortened and performed
external work or contracted isometrically. Oxygen uptake appeared
to be constant if load was constant and rest length changed. These
data disagree in effect with the preceding studies of oxygen uptake
and of energy exchange derived from myothermic or chemical measurements. Interpretation of the data is further complicated by the
pennate structure of the muscle which has in the past been blamed
for the odd data derived from this muscle compared to sartorius and
some other muscles [14]. The data stand. It is the interpretation
that is difficult.

In an attempt to determine whether the results from gastrocnemius were entirely or in part due to the pennate structure of the
gastrocnemius, a parallel-fibered muscle had to be found on which

the same studies could be done. Sartorius and gracilis were found
to be unsatisfactory. Semitendinosis seemed excellent [22]. Muscle
oxygen uptake ranged around 0.6 µl/g twitch. The relationship
between oxygen uptake and load in afterloaded twitch contractions
showed increased oxygen uptake with load similar to that seen in
gastrocnemius. This relationship was not true for rest lengths
beyond optimal length where oxygen uptake decreased and the contraction strength became unsteady. When load was held constant and
rest length changed, oxygen uptake was maximal at optimal length
and decreased when rest length was either longer or shorter. It
would appear that gastrocnemius and semitendinosis differ in these
types of contractions. However, it must be kept in mind that the
rest length changes in gastrocnemius were much less than those in
semitendinosis. As a result the changes related to rest length
alteration may not have been large enough to be detectable.

    If the data beyond rest length are ignored for simplicity then
the constant load experiments on semitendinosis muscles show a
direct relation between oxygen uptake and rest length, shortening
and external work. The constant rest length experiments show oxygen
uptake increasing with load. Because external work is greatest at
middle loads, the relation between oxygen uptake and external work
is curved, beginning on the ordinate with low oxygen uptake at zero
load, and extends counter clockwise to maximal work and back to the
ordinate at the highest oxygen uptake where the contractions are
isometric.

V. DISCUSSION

    The results described by Fischer [10] and Whalen and Collins
[24] differ mainly at the highest loads where the contractions were
isometric. Fischer found the oxygen uptake under these conditions
to be lower than the oxygen uptake at a lower load where external
work was done. Whalen and Collins [23] did not find this decrease
in oxygen uptake for isometric contractions. We agree with Whalen

and Collins [24] that the difference is related to the difference
in temperatures at which the experiments were done. The effect of
temperature on muscle contraction energetics has been shown clearly
by Fischer [11] where it seems apparent that at lower temperatures
the energy exchange for isometric contractions decreases relative
to that at lower loads. Since we have no new information about
this phenomenon and can not explain it, in the discussion below we
consider the relationship only as it appears at higher temperatures
where the largest oxygen uptake is for isometric or nearly isometric
contractions.

Another observation which appears in this review is that the
extra oxygen uptake for a contraction appears remarkably the same
for different muscles, muscles from different species, and for
different temperatures. Frog skeletal muscle oxygen uptake at
temperatures from 8 to 27.5°C are similar. Frog and dog skeletal
muscles are similar as are dog heart and skeletal muscle, the range
being from 0.1 to 1.0 $\mu l/O_2/g$ twitch if Baskin's data [1,3] are
excluded.

The oxygen uptake data from the mammalian muscles, heart and
skeletal, seem similar and they in turn are similar to the data
from frog sartorius at higher temperatures. The basic relationship
for the three muscle types is that between oxygen uptake and load
for afterloaded contractions. Oxygen uptake per contraction increases with load but at a decreasing rate as load increases, the
increase in oxygen uptake being quite small at the highest loads as
the contractions approached being isometric. This relationship has
been seen also in heart muscle when the energetic relationships are
investigated by chemical measurements [17] and by heat measurements
[12]. We believe that, however measured, this is the true relationship for energy exchange and load for such afterloaded contractions.

A fundamental question is why the relationship should be like
this. It has been suggested that the energy exchange is only
related to tension development [15,21,23] or to contractile element

work [4,6]. It has also been suggested that it is due to the
combined effects of several mechanical factors which add algebra-
ically [12].

Different sources of data presented suggest that tension
development or tension-time are not likely to be the only factors.
For example, oxygen uptake increased with increasing rest length,
shortening and external work in the constant load experiments on
semitendinosis muscle [22]. In addition it has been observed that
in afterloaded contractions oxygen uptake can be constant when load
is increased if with each increase in load, rest length is decreased
slightly (unpublished observations). In fact it is possible to
question whether tension has been proven to have any role at all.
For instance, when most mechanical variables are held constant, as
in isometric contractions, the tension development is altered by
changing rest length. The increase in tension is associated directly
with the change in rest length. Perhaps rest length is the important
factor. Certainly tension is not the only factor.

Gibbs and Gibson [12] have suggested that rest length by itself
is not a factor. Their argument stems from referenced data of their
own and from the observation that increasing stroke volume in the
in situ heart, with constant blood pressure, requires increased
rest length, but was not accompanied by an increase in heart oxygen
uptake [18,19]. These latter data also suggest that shortening and
external work have no effect on oxygen uptake. We think these in
situ heart as a pump studies cannot be interpreted so easily into
the functional terms used for skeletal muscles pulling on a good
myograph. As a result we suggest these data may not be applicable
to skeletal muscle and do not prove rest length is not a factor.
Gibbs and Gibson's other data in this regard are largely circumstan-
tial. In direct contrast to Gibbs and Gibson's suggestion, the
semitendinosis muscle of the dog showed a large increase in oxygen
uptake with rest length increase under conditions of constant load.
Such contractions seem analagous to increasing stroke volume at
constant pressure, but the results are quite different. Rest length

remains a possible factor, but like tension, it cannot account for
the energy exchange as the only factor.

Britman and Levine [4] and Coleman et al. [6] have suggested
that contractile element work is the only determinant of myocardial
oxygen uptake. The data of Britman and Levine [4] are for intact
heart and difficult to interpret. Those of Coleman et al. [6] are
on isolated papillary muscle strips, a complicated and sophisticated
study. Their calculation of internal work required estimation of
series elasticity. The assumption is made that all the internal
work is done on this component. In addition they measured external
work with a muscle lever. The report shows two different energetic
relationships for contractile element work. One is for oxygen
uptake related to internal work and another is for oxygen uptake
related to external work. Presumably contractile element work is
the force through a distance product summed for all the sarcomeres.
Why or how sarcomere work can be related differently depending upon
whether the sarcomere work is done on elastic tissue or a lever is
not questioned in the study. We therefore question in turn the
meaning of the measurement of internal work and its relation to
oxygen uptake. If internal work is questionable in its relationship
to energy exchange, then external work can be questioned also relative to its role in determining energy exchange.

There is no other unequivocal evidence that work per se,
internal or external, has a direct relationship with energy exchange.
The gastrocnemius and semitendinosis muscles show the same relationship between oxygen uptake and external work for afterloaded isotonic
contractions with increasing loads. The quantitative relationships
are also alike. The maximal work done per gram of muscle is the
same for the two muscles even though gastrocnemius shortened little
and had a high tension and semitendinosis muscle pulled less hard,
but shortened farther. Each showed a maximal efficiency for external
work of about 30%. However, gastrocnemius showed the same oxygen
uptake for a given tendon tension whether the muscle shortened and
did work or contracted isometrically. In contrast to gastrocnemius,

semitendinosis showed a lower oxygen uptake for a given tendon tension when the contractions were isometric. The difference was possibly due to the changes in rest length necessary to produce isometric contractions with the same tendon tension as the isotonic contractions. It has been suggested that gastrocnemius rest length could not be changed enough to uncover this variable. Nevertheless, the role of work, external or internal, is not clear.

The basic problem in studies like those described here is that one variable cannot be changed at a time. As a result, the inter-actions of the variables are difficult to separate and the effect of each cannot be determined. Our present interpretation would suggest that tension and rest length are the factors of importance in determining energy exchange. It is possible that, below optimal length, shortening may be a factor also. This interpretation is based more on negative data than positive data, and is not réally a forward step. Some sort of new information is needed.

The fact that tension changes produce a staircase-like effect in the intact heart [20] and the lack of clear relationship of oxygen uptake to both tension and rest length under a variety of circumstances in skeletal muscle and in intact heart, suggest to us another possibility. It may be that the variables described do not act in each single contraction to establish the amount of energy utilization. Instead these variables may act indirectly to alter the activation process which in turn determines energy exchange for subsequent contractions. This possibility is largely without precedent, but a new approach seems needed.

REFERENCES

[1] Baskin, R. J., "The variation of muscle oxygen consumption with load," J. Physiol. (London), 181, 270-281 (1965).

[2] Baskin, R. J., "The variation of muscle oxygen consumption with velocity of shortening," J. Gen. Physiol., 49, 9-15 (1965-1966).

[3] Baskin, R. J., and S. Gaffin, "Oxygen consumption in sartorius muscle," J. Cell. Comp. Physiol., 65, 19-25 (1965).

[4] Britman, N., and H. J. Levine, "Contractile element work: a major determinant of myocardial oxygen consumption," J. Clin. Investigation, 43, 1397-1408 (1964).

[5] Coleman, H. N., "Effect of alterations in shortening and external work on oxygen consumption of cat papillary muscle," Am. J. Physiol., 214, 100-106 (1968).

[6] Coleman, H. N., E. H. Sonnenblick, and E. Braunwald, "Myocardial oxygen consumption associated with external work: the Fenn effect," Am. J. Physiol., 217, 291-296 (1969).

[7] DiPrampero, P. E., P. Ceretelli, and J. Piiper, "Energy cost of isotonic tetanic contractions of varied force and duration in mammalian skeletal muscle," Arch. Ges. Physiol., 305, 279-291 (1969).

[8] Fales, J. T., S. R. Heisey, and K. L. Zierler, "Dependency of oxygen consumption of skeletal muscle on number of stimuli during work in the dog," Am. J. Physiol., 198, 1333-1342 (1960).

[9] Fenn, W. O., "The gas exchange of isolated muscle during stimulation and recovery," Am. J. Physiol., 83, 309-322 (1927-1928).

[10] Fischer, E., "The oxygen-consumption of isolated muscles for isotonic and isometric twitches," Am. J. Physiol., 96, 78-88 (1931).

[11] Fischer, E., "Die Wärmebildung des Skelettmuskels bei direkter und indirekter Reizung, sowie bei der Reflexzuckung," Arch Ges. Physiol., 219, 514-553 (1928).

[12] Gibbs, C. L., and W. R. Gibson, "Energy production in cardiac isotonic contractions," J. Gen. Physiol., 56, 732-750 (1970).

[13] Lee, K. S., "The relation of the oxygen consumption to the contraction of the cat papillary muscle," J. Physiol. (London), 151, 186-201 (1960).

[14] Martin, D. S., "The relation between work performed and heat

liberated by gastrocnemius, semitendinosis, and tibialis anticus of the frog," Am. J. Physiol., 83, 543-547 (1927-1928).

[15] McDonald, R. A. Jr., "Developed tension: a major determinant of myocardial oxygen consumption," Am. J. Physiol., 210, 351-356 (1966).

[16] Mommaerts, W. F. H. M., "Energetics of muscular contraction," Physiol. Rev., 49, 427-508 (1969).

[17] Pool, P. E., B. M. Chandler, S. C. Seagren, and E. H. Sonnenblick, "Mechanochemistry of cardiac muscle. II. The isotonic contraction," Circulation Res., 22, 465-472 (1968).

[18] Sarnoff, S. J., E. Braunwald, G. H. Welch, Jr., R. B. Case, W. N. Stainsby, and R. Macruz, "Hemodynamic determinants of oxygen consumption of the heart with special reference to tension-time index," Am. J. Physiol., 192, 148-156 (1958).

[19] Sarnoff, S. J., J. P. Gilmore, N. S. Skinner, A. G. Wallace, and J. H. Mitchell, "Relation between coronary blood flow and myocardial blood flow," Circulation Res., 13, 514-521 (1963).

[20] Sarnoff, S. J., J. H. Mitchell, J. P. Gilmore, and J. P. Remensnyder, "Homeometric autoregulation in the heart," Circulation Res., 8, 1077-1091 (1960).

[21] Stainsby, W. N., "Oxygen uptake for isotonic and isometric twitch contractions of dog skeletal muscle in situ," Am. J. Physiol., 219, 435-439 (1970).

[22] Stainsby, W. N., and J. K. Barclay, "Relation of load, rest length, work and shortening to oxygen uptake by in situ dog semi-tendinosis," Am. J. Physiol., 221, 1238-1242 (1971).

[23] Whalen, W. J., "The relation of work and oxygen in isolated strips of cat and rat myocardium," J. Physiol. (London), 157, 1-17 (1961).

[24] Whalen, W. J., and L. C. Collins, "Work and oxygen consumption in the frog sartorius muscle," J. Cell Comp. Physiol., 61, 293-299 (1963).

Numbers in brackets are reference numbers and indicate that an author's work is referred to although his name is not cited in the text. Underlined numbers give the page on which the complete reference is listed.

A

Abbott, B.C., 73, 86
Abe, Y., 125[1], 142
Abrahams, V.C., 139, 142
Adal, M., 152[1,17], 155[1,17], 157, 164[1], 165[1], 167[17] 169[17], 170[2,17], 172[1], 211, 212
Adams, R.D., 188[3], 189[3], 211, 256, 269
Afifi, A.K., 245[71], 253
Agarwal, G.C., 191[72], 217
Ahlquist, R.P., 120[3], 142
Albers, R.W., 183[79], 218, 224
Allbrook, D.B., 15[2], 20, 38, 256, 269
Allison, J.B., 109[44], 111[44], 117
Aloisi, M., 14[15], 26, 39
Alpert, N., 92[1], 115
Anderson, H.K., 131[23], 144
Andersson-Cedergren, E., 175[96], 177, 179[98], 211, 219
Andres, K.H., 152[59], 155[59], 157[59], 170[59], 216
Andrew, B.L., 148[5], 211
Angers, D., 191[6], 211
Annoni, J.D., 29[73], 44
Arvill, A., 100, 115
Ashford, T.P., 94, 117
Askansas, V., 265[28], 269[3], 269, 272
Asmundson, V.S., 255[4], 261, 265[20], 269, 271

B

Bajusz, E., 234[27], 244[27], 249
Baker, W. de C., 15[2], 20, 38
Barany, M., 65[1], 66, 68
Barclay, J.K., 280[22], 282[22], 286
Barker, D., 148[8], 149[9], 151, 152[17], 155[17], 167, 169, 170[2], 175, 176, 181[12], 183[9], 187[7], 211, 212
Baron, J., 92[30], 94[30], 116
Bartsoua, D., 94, 115
Baskin, R.J., 275, 276, 281, 284, 285
Batten, F., 187[18], 188, 212
Beilin, R.L., 180[159], 181[158], 225
Bell, C.D., 259, 261[5], 269
Benson, R. W., 54[16], 69
Bessou, P., 207, 212
Bickerstaff, E.R., 241[55], 251
Bigland, R., 95[5], 115
Blomfield, L.B., 16[32], 40
Blumberg, J.M., 256[27], 272
Bock, W.J., 17[25], 40
Bourne, G.H., 180[131], 222, 255[6], 269
Bowen, T.E., Jr., 73[12], 79[12], 84[8], 86, 87
Bower, E.A., 136[4], 142
Boyd, I.A., 149[22], 151[24], 155[26], 167[24], 169, 187[22], 189, 192, 202, 207[25], 212, 213

Striated muscle, oxygen uptake
    by, 273-286
  in frogs, 274-276
  in mammals, 276-278

                    T

Tendons, role of connections of,
    in muscle regeneration, 23-24
Tension, role in muscle regener-
    ation, 24-26
Tortoise, intrafusal muscle
    fibers of, 172-174

                    U

Uterus, adrenergic transmitter
    effects on, 119-146

                    V

Vascular supply, of muscle,
    re-establishment after
    regeneration, 19-21